Case Studies
in
Mathematical Modelling

Case Studies
in
Mathematical
Modelling

edited by

D. J. G. James, BSc, PhD, FIMA

Head of Mathematics Department, Coventry (Lanchester) Polytechnic,
Coventry, England

and

J. J. McDonald, BSc, MSc

Senior Lecturer in Mathematics, Department of Mathematics and
Computing, Paisley College of Technology, Paisley, Scotland

A HALSTED PRESS BOOK

John Wiley & Sons
New York

First published in 1981 by:
Stanley Thornes (Publishers) Ltd, Cheltenham, England

Published in the U.S.A. by:
Halsted Press, a Division of John Wiley & Sons, Inc., New York

ISBN: 0 470 27177-9

Typeset and printed in Great Britain.

78125710

CONTENTS

References are given at the end of each chapter.

PREFACE

This book is the result of a Workshop on the Teaching of Mathematical Modelling organised jointly by the Mathematics Departments of Coventry (Lanchester) Polytechnic and Paisley College of Technology. The motivation for organising such a workshop was that both institutions were involved in teaching modelling skills on their undergraduate courses and were finding the task far from easy. The difficulties arose from many directions; for example, the uncertainty as to what and how to teach in this field and, perhaps most of all, lack of good readily available source material. The main objective of the workshop, therefore, was to bring together people active in the teaching and use of mathematical modelling in order to prepare source material. The following abstract from the background information sent to potential participants is worth repeating here, in that it summarises the viewpoints of the organisers prior to the workshop.

> In recent years there has been a growing realisation that the subject of mathematical modelling is one which has been neglected in undergraduate courses. McLone,[1] for example, has collected statistics relevant to the education of mathematicians in the UK. These statistics confirm that mathematicians and their employers rank skills in problem formulation and communication at least as highly as ability in solving mathematical problems. In the USA, Gaskell and Klamkin[2] have carried out a more qualitative study of how heads of non-mathematics groups view the training of mathematicians. The views expressed in this study also indicate that recent graduates were weak in applying mathematics to real problems and communicating results. Our own experience has not led us to question McLone's summary of the average mathematics graduate:
>
> Good at solving problems, not so good at formulating them, the graduate has a reasonable knowledge of literature and technique, he has some ingenuity and is capable of seeking out further knowledge. On the other hand, the graduate is not good at planning his work, nor of making a critical evaluation of it when completed; and in any event he has to keep his work to himself as he has apparently little idea of how to communicate it to others.
>
> For the past few years we have been attempting to improve our own courses by introducing the teaching of mathematical modelling at an early stage.[3,4] We have made a conscious effort to concentrate on the presentation of 'modelling methodology'. The latter is the approach which a mathematician uses in solving practical problems and may therefore be idiosyncratic. There is, however, enough common ground for the main stages of the process to be stated and taught.
>
> Although it is rarely possible to confront junior students with real situations and systems, we have tried to prepare material involving situations which, whilst relatively simple, have sufficient realism to highlight all of the stages of the modelling process. One example of this material is given in reference 5.

Our aim whilst teaching modelling methodology has been to avoid, as far as possible, the introduction of new techniques of mathematical analysis. In trying to do this, we have been frustrated by the lack of suitable published material at the appropriate level. Titles such as in references 6-8 at first sight look promising, but on closer inspection are found to be vehicles for introducing techniques of mathematical analysis in a practical context. The concept of the M.A.A. workshop[9] excited us, but the published papers did not place the emphasis where we had hoped to see it.

In planning the Workshop on the Teaching of Mathematical Modelling we intend to place the emphasis on understanding real systems, formulating problems and interpreting the results of the analysis. The amount of mathematical analysis required would be appropriate to the stage at which the material would be presented and the methods would be already known to the students.

1. R.R. McLone, *The Training of Mathematicians*, SSRC Report, 1973.

2. R.E. Gaskell and M.S. Klamkin, The Industrial Mathematician Views his Profession, *Am. Math. Monthly* **81**, 699-716 (1974).

3. J.J. McDonald, Introducing Mathematical Modelling to Undergraduates, *Int. J. Math. Educ. Sci. Technol.* 8(4), 453-461 (1977).

4. D.J.G. James and N.C. Steele, Mathematical Modelling in Undergraduate Courses, *Proc. A.A.M.T. 8th Biennial Conf.*, Canberra, 1980.

5. J. Huntley and J.J. McDonald, Buying a Car—An Example of Mathematical Modelling Applied to Decision Making, *J. Math. Modelling for Teachers* 2(1), 14-24 (1979).

6. P. Lancaster, *Mathematics, Models of the Real World*, Prentice Hall, 1976.

7. R. Haberman, *Mathematical Models*, Prentice Hall, 1976.

8. J.G. Andrews and R.R. McLone, *Mathematical Modelling*, Butterworths, 1976.

9. M.A.A., *Modules in Applied Mathematics*, Mathematical Association of America, 1976.

The Workshop was organised in the form of two three-day sessions. The aim of the first session was to establish objectives and guidelines for the preparation of material to be presented at the second session. Consequently, the first session was arranged so that participants could present their views on the nature of mathematical modelling and the teaching of the subject. These presentations were then followed by group discussions on the nature of the material to be prepared for the second session. From these group discussions there evolved the basic ideas from which a number of case studies on the teaching of mathematical modelling were developed. Drafts of these case studies were discussed at the second session and since then have been subjected to revision as appropriate.

The workshop participants do not regard the case studies as a course in mathematical modelling. Rather they see them as a source which those teaching mathematical modelling, or the application of mathematics, may draw on in whatever way they consider most appropriate. Although the approach taken in each of the studies may be somewhat idiosyncratic (for that seems to be the nature of mathematical modelling) readers will find that each has been structured in much the same way.

Each study has a self-contained statement of the problem to be considered and this, in many cases, is accompanied by other material providing background and motivation. It is envisaged that lecturers/teachers using the case studies within the teaching context, will direct the students in the first instance only to this introductory material. As guidance the relevant sections for each case study are listed in the section immediately following this preface. Participants at the workshop were strongly of the opinion that students should not, at the outset, by given information on how to approach the particular problem; rather this should evolve in later discussions.

In developing an appropriate model the author of each case study will have adopted a particular stance. The approach chosen will be based on the author's use of the study with a group of students and, for this reason, authors have provided comments and hints on handling the study in a teaching environment. The workshop participants were conscious of the danger of appearing dogmatic, but on balance, it was felt that there were instances where the author's advice, based on experience, could be valuable to a teacher or student. Sections which are included to provide advice are indicated by a bold line in the margin. However, it cannot be emphasised enough that there is no one 'answer' to any of the problems contained in this collection; others may take quite different approaches to that of the authors.

In the first chapter we deal very briefly with the nature and purpose of models and with the methodology of model building. Prior to the first session of the workshop some of the participants felt that it would be possible to identify and describe an approach to model building which was sufficiently unified and structured to be described as a methodology. During the session, however, it became apparent that the approach to model building adopted by participants were quite varied. This first chapter, therefore, is not intended to be seen as anything more than a collection of observations which might place the individual studies in a suitable context and which will be of use to the student meeting mathematical modelling for the first time.

The case studies forming the remainder of the book may be attempted in any order. The demands made on mathematical techniques vary from one case study to the other. A number can be tackled using pre-college mathematics; these include 'The Humber Tunnel Authority', 'Selling Hot Dogs', 'Modelling the True Cost of a Mortgage', and 'Air Gap Coiling of Steel Strip'. Others depend explicitly on certain areas of mathematics; for example, 'Describing Plant Growth', 'Hydroelectric Power Generation

System', and 'Crop Yield and Number of Plants', rely on some knowledge of differential equations, whilst 'Gas Leaks' and 'The News-Vendor's Dilemma' require a little knowledge of probability. A limited number, particularly 'Speed-Wobble in Motorcycles,' are more demanding in their use of mathematical and computational techniques.

As organisers of the workshop we would like to thank all the participants for their contribution and also express our gratitude to The Nuffield Foundation for their financial support. Thanks are also due to Adam and Charles Black Ltd. for permission to quote, in Chapter 1, from *The Pursuit of Stillwater Trout* by Brian Clarke and to Academic Press Inc. (London) Ltd. for permission to reproduce the table of Plant Processes and Properties (Table 5.1) from *Mathematical Models in Plant Physiology* by J.H.M. Thornley.

Finally we wish to express our gratitude to the staff of Stanley Thornes (Publishers) Limited for their cooperation and helpful suggestions in preparing the text for publication.

D.J.G. J.
J.J. McD.

CONTRIBUTORS

D.A. Abercrombie, BSc, PhD, AFIMA, MIS, FSS.
Systems Planning Engineer, Forward Planning Branch, South of Scotland
Electricity Board, Glasgow, Scotland.

A. Barnes BSc, MSc.
Statistician, National Vegetable Research Station, Wellesbourne, Warwick
CV35 9EF.

D.N. Burghes, BSc, PhD, FIMA, FRAS.
Director of the Cranfield Centre for Teacher Services, Cranfield Institute
of Technology, Cranfield, Bedford, MK43 0AL.

R. Croasdale, BSc, DIC, MSc, MIS.
Principal Lecturer, School of Mathematics, Statistics and Computing,
Newcastle-upon-Tyne Polytechnic.

M. Cross, BSc, PhD, FIMA, MMS.
Visiting Professor, Mineral Resources Research Centre, University of
Minnesota, USA.

S.C. Dunn, MSc, FIEE, C Eng.
Chief Research Engineer, British Aerospace, Dynamics Group, PO Box
19, Six Hills Way, Stevenage, Hertfordshire SG1 2DA.

I.D. Huntley, BA, PhD, FIMA.
Senior Lecturer in Mathematics, Department of Mathematics and Computing, Paisley College of Technology, High Street, Paisley, Renfrewshire,
PA1 2BE, Scotland.

D.J.G. James, BSc, PhD, FIMA.
Head of Mathematics Department, Coventry (Lanchester) Polytechnic,
Priory Street, Coventry CV1 5FB.

J.J. McDonald, BSc, MSc.
Senior Lecturer in Mathematics, Department of Mathematics and Computing, Paisley College of Technology, High Street, Paisley, Renfrewshire,
PA1 2BE, Scotland.

A.E. Millington, MA, AFIMA.
Lecturer in Mathematics, Department of Mathematics, Liverpool Polytechnic, Byrom Street, Liverpool, L3 3AF.

M.A. Wilson, MA, MSc, MBCS.
Senior Lecturer in Operational Research, Department of Statistics and Operational Research, Coventry (Lanchester) Polytechnic, Priory Street, Coventry CV1 5FB.

A.E. Moscardini, BSc, MSc.
Senior Lecturer in Mathematics, Department of Mathematics and Computer Studies, Sunderland Polytechnic, Priestman Building, Green Terrace, Sunderland SR1 3SD.

K.H. Oke, BSc, M Tech, FIMA.
Principal Lecturer in Mathematics, Department of Mathematical Sciences and Computing, Polytechnic of the South Bank, Borough Road, London, SE1 0AA.

A. Rothery, BSc, DIC, PhD, AFIMA.
Worcester College of Higher Education, Henwick Grove, Worcester, WR2 6AJ.

N.C. Steele, BSc, MSc, FIMA.
Senior Lecturer in Mathematics, Department of Mathematics, Coventry (Lanchester) Polytechnic, Priory Street, Coventry CV1 5FB.

J.R. Usher, BSc, PhD, AFIMA.
Senior Lecturer in Mathematics, Department of Mathematics, Glasgow College of Technology, Cowcaddens Road, Glasgow, G4 0BA, Scotland.

GUIDANCE TO USING THE CASE STUDIES

When tackling a particular case study, the user should, in the first instance, restrict attention to the introductory material only, which includes the statement of the problem. Consequently, a lecturer or teacher using the material for teaching purposes should initially direct the student's attention only to this introductory material. To assist in this exercise relevant sections for each case study are listed below.

1. **An Introduction to Modelling**

   This chapter contains some introductory remarks on models, the process of model building and the use of the case studies which form the remainder of the book.

2. **Crop Yield and Number of Plants**

   See Section 2.3 for the author's advice on how to handle this case study.

3. **Deer Harvesting**

   Sections 3.1 and 3.2.

4. **Modelling the Leaching of Nitrates in Fallow Soils**

   Sections 4.1 to 4.3.

5. **Describing Plant Growth**

   Sections 5.1 and 5.3.

6. **Tumour Growth and the Response of Cells to Irradiation**

   This chapter contains two case studies which may be studied independently. The relevant introductory sections for each case study being:

   Tumour Growth: Sections 6.2, 6.3.1 and 6.3.2.

   Response of Cells to Irradiation: Sections 6.2, 6.4.1 and 6.4.2.

7. **The Humber Tunnel Authority**

   Section 7.1.

8. **Parking a Car**

   This case study is presented in a somewhat different format to the others and is included as an interesting article written by an industrial mathematician. As well as giving an insight to a novel practical approach the author poses a number of problems which a lecturer or teacher could develop into individual case studies.

9. **Air Gap Coiling of Steel Strip**

   Sections 9.1 and 9.2. See also Section 9.7 for author's advice on how to handle this case study.

10. **Hydroelectric Power Generation System**

    Section 10.2.

11. **Speed-Wobble in Motorcycles**

    See Section 11.3 for author's advice on how to handle this case study.

12. **The News-Vendor's Dilemma**

    Initially only Section 12.1 and subsequently the material of Section 12.4.2.

13. **Modelling the True Cost of a Mortgage**

    Section 13.1

14. **Searching for Gas Leaks**

    Initially Section 14.1 and subsequently the data of Section 14.4.

15. **Selling Hot Dogs**

    Section 15.1.

16. **Product Pricing**

    Section 16.1.

1.

An Introduction to Modelling

1.1 MODELS AND MODELLING

It seems to be the done thing, when introducing a topic in mathematics, to begin with definitions. It is particularly difficult to give a brief and comprehensive definition of a mathematical model. First let us look at a few dictionary definitions of the word 'model' to assess how the common use of the word relates to its mathematical usage.

- Representation in three dimensions of a proposed structure.[1]

- Design, style of structure.[1]

- Person, thing, proposed for imitation.[1]

- A set of plans for a building to be erected or of drawings to scale for a structure already built.[2]

- A, usually three-dimensional, representation of something existing in nature or constructed or to be constructed.[2]

- A collection of statistical data; or an analogy used to help visualise, often in a simplified way, something that cannot be directly observed (as an atom).[2]

- A simplified representation or description of a system or complex entity especially one designed to facilitate calculations and predictions.[3]

It is rather difficult to find a concise, useful definition of the word 'model' which makes some concession to the world of mathematics. Even quite modern dictionaries of mathematical terms seem to ignore the subject. The word is relatively new in mathematics and would seem to have been borrowed from earlier usage because of the analogy between the mathematical model and the scale or working model. As will be appreciated later, the former can be used just as much as the latter as a means of experimentation. It is gratifying to note that mathematics, in its turn, is making an impact on normal English usage in that the last two of the definitions above have very clear mathematical origins.

In order to get a feel for what the words 'models' and 'modelling' mean in a mathematical context let us look at an example. A common problem experienced by companies involved in the distribution of goods by road is ensuring that the goods are delivered to the right place at the right time but without excessive use of resources. To achieve this end, a considerable amount of time is spent by these companies in route planning — that is, in deciding which vehicles should deliver which goods to which customer and the order and time at which these deliveries should be made. Another aspect of the problem may be deciding on the number and location of warehouses and the number and types of vehicles required. Very often routes have to be planned on a daily basis and have to take into account such matters as the demands of the customers, the capacities of the vehicles, statutory or other restrictions on the length of time which the drivers may work, limited times of access to customers' premises, and the length, width and height restrictions on vehicles at certain premises. For a small company with only a few vehicles, route planning may well be done by an experienced route planner working by trial and error but backed up by plenty of practical know-how. For a larger company, with perhaps twenty to thirty vehicles of different types and several hundred customers, the problem can be quite horrendous and it may also have to be solved each day. Consequently several companies market computer packages which can be used for route planning. At the heart of such a package there is usually a model of the problem on which calculations are carried out to help in the planning process. In the bad old days, after working through an analysis of the model, a computer would produce the definitive routes. In many cases the solution produced was not really acceptable to an experienced planner who could easily find faults in it. This, for some time, gave the model-oriented approach a bad name which should be taken as a warning in that modelling should not be carried out in isolation. Fortunately, the latest computer packages, although using similar models, make use of computer graphic aids and allow the human planner to interact with the model to produce a more acceptable solution.

In order to exemplify some of the key features of models and the modelling process let us look at one of the early route planning (or vehicles scheduling) models. First we consider a formulation of the model.

A set of customers of known location and with known requirements for a certain commodity is to be supplied from a single depot. The problem is to schedule a number of vehicles to make these deliveries subject to the following constraints.

(a) Each vehicle has a load or capacity limitation which cannot be exceeded.

(b) There is a limit on the maximum trip length of a vehicle. This is related to statutory restrictions on the length of driving periods.

Readers will note that this information lacks some of the generality of the problem first stated. Suppose that you, the reader, as the company mathematician or consultant, had been given this problem. Then it may have been given to you in one of the following forms:

The last oil price increase made us realise that we are using far too much fuel in the transport division. Is there any way in which we can reduce our fuel bill?

or

We think we have more vehicles than we require. Three vehicles are due for replacement within the next few months. Can we avoid replacing them by using the remaining vehicles more efficiently?

After carrying out an initial investigation and giving the problem some thought you may have decided that the above formulation is adequate for considering the aspects of the problem you wish to consider at the moment. You may realise that there are other aspects of the problem, such as restrictions on access, which you will return to later once you have solved the core of the problem.

The important point to be emphasised here is that the formulation of the problem is an economical statement, usually in English, of the aspects of the problem to which the model must relate. The mathematical model should bear a close correspondence to the formulation. With this in mind, a model relating to the above formulation can now be given. There is no attempt here to indicate how such a model was developed, as this is the purpose of most of the case studies in this book. The aim here is to show what a typical model looks like. The reader should merely note that the mathematical model is a set of equations which is closely related to the formulation of the problem.

The variables used have the following meanings:

n = the number of customers

q_k = the requirement of customer k

Q = the vehicle capacity

c_{ij} = the cost of the trip from i to j

y_{ijk} = the quantity shipped from i to j destined for k

x_{ij} = 1 if a vehicle visits j from i

= 0 otherwise

T = total cost of the vehicle trips,

where the subscripts can take on the values $k = 1, 2, \ldots, n$; $i = 0, 1, 2, \ldots, n$; $j = 0, 1, 2, \ldots, n$ with 0 referring to the depot.

The model consists of the following relationships between the variables.

$$T = \sum_i \sum_j c_{ij} x_{ij} \qquad [1.1]$$

$$\sum_i y_{ijk} = \sum_r y_{jrk} \quad \text{for all } j, k \ (j \neq k) \text{ where } r \text{ is arbitrary} \qquad [1.2]$$

$$\sum_i y_{ikk} = q_k \quad \text{for all } k \qquad [1.3]$$

$$\sum_k \sum_j y_{0jk} = \sum_k q_k \qquad [1.4]$$

$$\sum_i x_{ij} = \sum_r x_{jr} = 1 \quad \text{for all } j \qquad [1.5]$$

$$\sum_k y_{ijk} < x_{ij} Q \quad \text{for all } i, j \ (i \neq j) \qquad [1.6]$$

Relationship [1.1] expresses the total cost of deliveries as the sum of the costs of the individual links which are made in the delivery network, while relationship [1.2] expresses the fact that there is no accumulation of goods at j which are required at customer k. The reader may wish to interpret the other relationships, but the important thing at this stage is to appreciate that a mathematical model is just a set of mathematical relationships which reflect accurately the formulation of the problem. Of course the ultimate value of the model depends on both the extent to which it is capable of efficient analysis and its relevance to the real situation as originally posed. This early model was deficient in both respects.

The analysis of the model involves finding the links x_{ij} and the consequent routes which, subject to the constraints [1.2]–[1.6], minimise the cost T. This kind of problem is classified as a 'zero–one integer programming problem'. It is not important here that readers know how to solve problems of this type, only to appreciate that they normally have to be solved on a computer.

A disadvantage of this otherwise rather elegant model is that, for a realistic problem, there would be too many variables and equations for an economical solution to be carried out. For example, in a delivery situation with 200 customers there would be 40 000 zero–one (x_{ij})

variables. The consequence of this was that researchers sought other model structures to fit the basic formulation. They were mainly successful, in that models, and corresponding computational schemes, were found which could be used to solve the problem efficiently; but the quality of solutions, however, was still lacking. One of the deficiencies was that, in solutions which minimised total cost as described in the model, most vehicles were heavily utilised in terms of load and length of time spent working whilst a few were very lightly utilised. These solutions were not popular with drivers or route planners and the search was on again for a better approach. Notice that in this case it was not the mathematical structure of the model or the method of solution which was at fault, but the basic concept of minimising mileage costs.

The non-trivial situation here helps to put models and modelling within the general context of mathematics and its applications, and a number of points are worth noting.

(a) Professional mathematicians are seldom given mathematical problems 'on a plate'. They often have to create the mathematical problem.

(b) The mathematical problem (or model) is created through a sometimes lengthy process of formulation.

(c) Although the model is most closely associated with the problem formulation, it must be judged in terms of (i) the reliability with which it reflects the original problem and (ii) the efficiency with which it may be analysed.

(d) There is not necessarily a unique model for a particular situation. Even for a given formulation there may be more than one mathematical structure.

(e) To build good models and make an accurate assessment of their value it is necessary to be thoroughly competent in mathematical theory and techniques — and a lot more besides. Initiative and skills in communication are particularly important.

(f) Modelling is perhaps easier to define than model. It could be defined as the process of translating a problem from its real environment to a mathematical environment, in which it is more conveniently studied, and then back again.

1.2 MODELLING, MATHEMATICS AND MATHEMATICAL EDUCATION

Mathematical modelling, therefore, is an integral part of the practice of mathematics; indeed there is much in the history of the development of mathematics to support this view. Mathematics probably had its beginnings around 5000 years ago in the Middle East. The earliest crude forms of mathematics were developed by the Babylonians and the Egyptians as practical methods for helping them with everyday problems. These problems included keeping track of the days and seasons so that seeds could be planted at the right time, counting animals, and apportioning land and taxes. Mathematicians were particularly fascinated by the movement of the heavenly bodies and many models of the solar system — mathematical and physical — were proposed to explain the observed regular movements of the planets. Many of these, often complex, models explained some aspects of the observations; but it was not until the time of Copernicus, Galileo and Kepler in the Europe of the Middle Ages that the model of the solar system which we have come to accept took form. It was in attempting to handle motion and gravitation mathematically that Newton devised and used the calculus. It was in attempting to correct the inadequate descriptions, given by Newtonian mechanics, of energy and objects which move at nearly the speed of light that Einstein, at the start of this century, put forward the theory of relativity. In between the times of Newton and Einstein mathematicians such as Lagrange and Laplace developed theories, techniques and models which have been used by generations of engineers and scientists.

Clearly then, many of the important developments in mathematics were consequences of practical requirements and involved what would now be called models: models of a body falling freely under gravity, models of the motion of projectiles, models of the motions of planets, and so on. Alongside these immediately practical uses of mathematics there were, of course, many theoretical advances. These were led by the Greeks, who were the first to formulate the two mental processes necessary for progress in mathematics — abstraction and proof. Abstraction is the art of identifying the common features of different things, so that general ideas can be created. Proof is the art of arguing from premises to a conclusion in such a way that there is no flaw in the argument.

The spectrum of mathematical theory, techniques and applications has now become so extensive that no student of mathematics can hope to touch on all areas, and professional mathematicians often find themselves specialising in a particular area: Statistics, Fluid Dynamics, Control Theory, Numerical Analysis, and so on. Given the amount of material which has to be covered in an undergraduate mathematics course, it is more and more the case that students can study mathematics in isolation. Even when applications are studied within the mathematics curriculum, they will often be limited in scope and determined by the particular area of specialism of the College or University concerned. This contrasts with the situation prior to the 1950s when most students of mathematics would take the subject along with another discipline — often one of the physical sciences. Whereas it is quite possible for modern students of mathematics to study the subject in relative isolation from other subjects without questioning the relevance of what they are doing or even suspecting that the mathematical structures with which they are dealing have any practical significance, earlier generations of students were often trained in the investigative approaches of other disciplines (whether these be philosophy or the physical sciences).

This trend has not been without its effects, and the study, in the early 1970s by McLone,[4] of recently qualified mathematics graduates in industry has now become part of the folklore of those wishing to improve the effectiveness of mathematical education for applied mathematicians. McLone's summary of the employers' view of the average mathematics graduate is worthy of repetition here:

> Good at solving problems, not so good at formulating them, the graduate has reasonable knowledge of literature and technique, he has some ingenuity and is capable of seeking out further knowledge. On the other hand, the graduate is not good at planning his work or making a critical evaluation of it when completed; and in any event he has to keep his work to himself as he has apparently little idea of how to communicate it to others.

It was also about this time that Professor Lighthill[5] chose communication and the interaction of mathematics with other subjects and the real world as the theme of his Presidential Address to the Second International Congress on Mathematical Education.

These well publicised views were in sympathy with those of many people working at the applied end of mathematical education, and treatment of mathematical modelling was seen as one way in which the quality of education in applied mathematics could be improved. It was recognised by most of these people that mathematical modelling was not a new discovery or even a new vogue, but simply an area of mathematical education which had been neglected.

The authors of the case studies in this collection have all, in their individual ways, been involved in teaching mathematical modelling over the last few years. Most have been involved in education and practice. Probably no two authors will agree entirely on how mathematical models should be built, nor on the extent to which mathematical modelling should appear in courses or how the subject should be taught. What they are all agreed on is that, at the present time, students and teachers lack suitable case study material. They have each presented a problem situation and described how they would develop a model which would be of value in investigating that situation. Each documentation follows the same format, and hints on the presentation of the case study are included where appropriate.

1.3 WHY BUILD MODELS?

The reader, by now, will probably have at least a hazy notion of why one would want to develop a mathematical model for a given situation. The following extract from the introduction to a book on trout fishing — not at all concerned with mathematics or modelling — probably conveys better than any mathematical text, the motivation which many people have on starting the development of a model. In it the author, Brian Clarke, describes his motivation for studying trout and for writing the book:

> For a long time, I thought that the difference between the expert angler, and the inexpert, was that the former possessed knowledge, whereas the latter did not...
>
> I suppose I took some minor, perverse consolation in the fact that I was not alone: in the realisation; indeed, that the great majority of all anglers were, like myself, modest performers at the waterside; and that their problems — the reasons that they performed as modestly as they did, and had a minimal level of knowledge — were almost certainly similar to my own.
>
> But then I had a stroke of what for me, at least, seemed inspiration: I realised that the critical difference between the expert at anything and the inexpert, is not information at all but understanding. I came to see that the inexpert angler fails most of the time because his success depends upon meeting conditions which coincide with a fixed, and usually limited, range of mentally catalogued techniques; whereas the expert angler, because of his fundamental understanding of what he is trying to achieve, in relation to the fish he is after, thinks more in terms of how and why, than of what; and thus is able to devise specific techniques in response to the demands of specific conditions. Through understanding, as it were, he achieves a kind of infinite flexibility.
>
> The realisation that this was so, changed my own approach utterly. Up to that point, I had been reading books in a search for information, in a raw sense: information on tackle and techniques, and knots; information on casting, and flies, and fish. And so, however I looked at it, my problem seemed to be a need to commit to memory an impossible series of only tenuously-linked, separate pieces of fact and folklore. Now, I could see, that wasn't the requirement at all. The requirement was understanding, the first glimpses of which would presumably stem from a sensible rationale relating the most important of these, individual, separate pieces of information, to one another. If I could find that, or even something approaching it, I would be in a position to work out and meet the demands of changing conditions for myself, rather than to have to thumb through my memory to see what techniques it contained, and fail if there was nothing appropriate...

How often have businessmen been filled with the same kinds of thoughts? They have important decisions to make; they have lots of information and plenty of experience, but they can't quite relate the different bits of information, or there is a change in circumstances — a 2% increase in interest rates, say. Similarly, weather forecasters are faced with masses of information on temperature, atmospheric pressure and wind velocities at different heights and from locations all over the world. How do they relate this information and formulate a weather report?

The message here is not that mathematical models should be used in trout fishing, but that there are many situations in which we require a thorough understanding of the systems we are dealing with. Often the systems are complex and the effects of the interplay of different parts of the system are difficult to assess. It is in these circumstances that we can use mathematical models to gain a deeper understanding of the systems and to examine their reaction to changing conditions. Indeed, on reflection, it might just be possible to build a model which, given information on geographical location of the lake, time of year, time of day, lakeside vegetation, air temperature, water temperature, wind velocity and the like, would tell the angler what tackle to use — type of leader and fly, the technique to be used, and would forecast the total catch in a three-hour session. Perhaps that would take all of the fun out of the sport!

On a serious note, however, that little bit of fantasising introduces another use of mathematical models — the ability of a good model to forecast the behaviour of a system. There are probably three levels of sophistication at which mathematical models can be used. These are:

(a) in aiding our understanding of systems;
(b) in predicting the behaviour of systems at a future time or in changed circumstances; and
(c) in controlling systems.

Keeping on the fishing theme, but in a different environment, readers will probably recall that each year there are well publicised discussions in various quarters concerning the maintenance of fish stocks. These talks might take place between EEC fisheries' ministers on North Sea stocks or in the annual quota meeting of the International Whaling Commission. In the latter meeting, for example, member countries meet to decide how many whales of each species may be killed in the next season. Since it is generally recognised that whale populations have become dangerously deflated, there has been growing pressure to suspend whaling completely, at least for a number of years. Increasingly mathematical models — albeit fairly crude ones — have been used to provide information which can be used in setting quotas which will at least maintain the populations at their current levels. The first requirement of such a model is that it provides accurate information on the current population sizes. It is not possible to count whale populations, so the model uses data on reproductive rates, age structure, mortality rates and the return from known fishing effort, to estimate the population size. This is a field in which data have been meticulously maintained over many years. The model uses well-known techniques of population dynamics, and the simple concept that the lower the population the lower the return will be from a given fishing effort, to estimate actual population sizes. Used in this way, the model is at the first level of sophistication and merely aids understanding. In this kind of situation the scientists probably have the basic understanding, but the decisions which are taken are political decisions, and models can be used to provide the scientists with hard information to support their arguments. At this level the scientists will want to check that the model does support their basic understanding and vice versa.

Provided that the scientists do have confidence in the model, it can then be used in a predictive mode by, for example, predicting what will happen to the population if whaling continues unchecked or subject to various quotas.

Finally the model can be used to provide the information required to control the population — for example, the quotas required to maintain the population at its current level, or the period over which whaling would have to be suspended to allow stocks to recover to a higher level. Throughout this period of control the predictions of the model could be compared with sources of data on the population, as a continuing check on the performance of the model and the success of the management policies.

This is only one example; throughout the case studies in this book readers should be able to identify the contributions of the models to the situations described in the studies. In 'Parking a Car' for example, the model is used as a means of clarifying an understanding of a familiar mechanical system. It is only when a clear understanding of an existing system has been acquired that its deficiencies can be identified and new designs considered. In 'Modelling the True Cost of a Mortgage' a familiar personal decision-making situation is examined — whether to take out a conventional mortgage or an endowment insurance-linked mortgage. In a similar way to Brian Clarke, the angler, the couple involved can collect the necessary information on interest rates, premiums and tax relief for themselves, but the problem is just complex enough to require a model to help in the resolution of the various courses of action possible. The case study entitled the 'Humber Tunnel Authority' shows, in a simple way, how a mathematical model can be used in the control of the flow of traffic, and the study on 'Deer Harvesting' explores how mathematical models might be used to help wildlife management control deer populations.

In each case study the mathematical model will be seen in action in at least one of its three roles, and in some cases in all three.

1.4 MODELLING METHODOLOGY

Having examined the motivation for building mathematical models it is appropriate to discuss whether or not there is such a thing as a 'modelling methodology'. At the outset of the workshop which generated these case studies, some of the participants were of the opinion that the process of model building was sufficiently capable of definition to be described at some length and that such a description could be called a methodology. By the close of the workshop, however, it had become apparent that different people use different approaches to modelling and the approach sometimes depended on the context. Without doubt, general experience as a mathematician and experience in a particular application area play a great part, but there is a general attitude and approach to modelling which is almost certainly worth conveying to the novice.

It is fairly widely recognised that there are a number of major steps in building and using mathematical models. These can be stated as follows.

(a) Recognition that a problem exists

(b) Familiarisation with the system to be modelled

(c) Formulation of the problem

(d) Construction of the model

(e) Validation of the model

(f) Analysis of the model

(g) Interpretation of the results

(h) Implementation

(i) Monitoring of the system and model.

(a) In more cases than not, the original recognition that there is a problem to be solved is not done by the model builder. What the model builder — particularly the novice — must remember is that the person who recognises the problem may only be recognising the symptoms of the problem and may not, in fact, have a clearly defined problem. For example, a manufacturer recognises that there has been a drop in demand for its products, and attributes this failure to the aggressive marketing tactics of its competitors. In fact it turns out that the fall in demand is due to a persistent failure to meet agreed delivery dates, but is this failure due to faulty stock-holding policies or poor production control or lack of capital investment?

(b) Mathematicians engaged in modelling very often find themselves working with researchers from other disciplines. This may be because they are employed by an organisation as mathematicians or perhaps because they earn their living as professional consultants. In many cases — particularly as consultants — they will be required to deal with systems and problems which are quite unfamiliar to them. It is fatal in such circumstances to attempt to carry on their activities in a 'back room'. They must be prepared to discuss problems fully with their 'clients' and to research thoroughly the background to the problem by reading books, papers and reports. They must ultimately be able to translate the results of their analysis back into the original context, and present them in a form which is understandable to their clients — and that means using their clients' language and not mathematical language. They also must not make the mistake of expecting too much in terms of answers from their clients — if they knew all of the answers they would not have involved a mathematician in the first place!

(c) Through this process of familiarisation the mathematician and client together arrive at a formulation of the problem. At the start of the investigation a very wide problem may have to be refined into something more specific, as was the case with the vehicle scheduling study. Alternatively, the mathematician may be presented with something as vague as 'Our sales have fallen. Can you help us remedy the situation?' or even 'I was at a management course last week and heard someone extolling the virtues of models — do you think models can help us in any way?'

The case studies in this book have been written in such a way that emphasis is placed on these steps of familiarisation and formulation. Each case study contains a statement of the problem which really gives the problem in semi-formulated form. The gaps need to be filled in together by the teacher/lecturer and the students. Teachers should familiarise themselves with the whole problem area and do any necessary background reading. Students too should do as much familiarisation as possible, but they may find it rather more difficult to get access to relevant background material in addition to that provided by the authors. A successful method of dealing with case studies is for the teacher to play the role of the client and the student that of the model builder. The more discussion that can be generated, the better the study will go. At the end of one study and discussion period it should be possible to write down a formulation of the problem.

(d) From this stage onwards the construction and analysis of the model can proceed. One of the things that students of mathematics find difficult to accept is that there is no unique approach, and perhaps no unique solution, to a real world problem. It is quite possible, therefore, that a different formulation of the problem from that of the author is arrived at and, even with the same formulation, a different model constructed. Students should have the confidence to try their own approaches and compare them with the approaches described by the authors.

It is in the construction of models from the formulation stage onwards that opinions diverge, both on how to piece models together and how to teach modelling. This book was never intended as a course on modelling, but as a collection of studies which could be used to supplement courses in modelling or applied mathematics. Students should seek advice from their lecturers or teachers on this aspect of the process and again informal discussion will probably bring best results. Nevertheless, although the approaches of the authors of the various case studies differ, each has attempted to give reasons for the approaches taken in formulation and subsequent development of the models and careful reading of the studies will give students useful insights into how models are constructed.

(e) Interwoven with the construction of the model there is the necessity to test the validity of the model at various stages. There are essentially three types of validity testing.

(i) *Testing while the model is being built.* The model may well have been synthesised from various component parts. It may be possible to test out the basic components before the model is completed. For example, it may be possible to check functional relationships between variables and parameter values by using existing data or by carrying out experiments. In practice sophisticated statistical techniques might be required but this kind of testing is beyond the scope of this book.

(ii) *Testing of the model on the basis of the results it produces.* In many cases, particularly at the elementary level of our case studies, it is possible to check the final behaviour of the model against the known behaviour of the system it represents. Especially when the model is to be used to predict the behaviour of a system under new circumstances, it is important to check it — if at all possible — against the behaviour of the system under known circumstances.

(iii) *Testing by implementation.* This method of testing is closely tied up with stages (h) and (i) of the modelling process — implementation and monitoring. If this final means of testing is required it needs a considerable leap of faith on the part of the model builder and the client, because it involves taking action on the basis of the information given by the model and then observing whether the action produces the desired effects.

(f) The nature of the analysis of the model depends on the form which the model ultimately takes. It is the stage of the process which most students of mathematics will be happiest with since it is what they are most used to doing, mathematically speaking! The analysis may involve some pencil-and-paper manipulation but frequently, in real life, at least part of the analysis will be done on a computer. Many of the case studies in this collection can be done using pencil-and-paper methods but with some of the models the analysis is eased or can be carried out in more depth with the aid of a computer. It has been assumed in writing the studies that students already have at their disposal the necessary mathematical and computing skills. The participants in the workshop have generally found that, in teaching modelling, it is best to use only mathematical techniques which the students have already encountered. If the aim is to teach modelling it is not a good idea to teach techniques at the same time. This, however, does not preclude the use of the case study problems as illustrations of the application of techniques covered in mathematics courses, but they were not written predominantly with that in mind.

(g) In all of the studies emphasis is placed on the interpretation of the results of the analysis of the model. For the student this can be one of the most interesting and rewarding parts of the exercise, since it is at this stage one can — for the first time — assess how well the model matches up to the reality of the original situation. In some cases there will be data against which the model can be checked, but in other cases it will be more a case of checking the behaviour of the model against qualitative information or an intuitive expectation of how the model should behave. Students should be in no way disappointed if they discover that the model does not behave as expected. Sometimes this will be because the student's intuition has been at fault — more often it will be the model that is faulty, through an error in formulation, construction or analysis. Although the whole exercise has to be reappraised, by this time the student will be well immersed in the problem and should find the exercise of seeking out the faults a stimulating one.

(h) It is rarely the case, in the undergraduate educational world, that the stages of implementation and monitoring can be carried out.

(i) As a last stage in the study, students should prepare a brief, non-mathematical report, outlining the significance of the results of the analysis. In writing this students should be aware that this may be the only chance to sell themselves and the value of their work, since the details of their work will probably be relegated to an appendix.

1.5 SOME FINAL ADVICE TO THE TEACHER

Students should be encouraged to view a modelling framework, such as that described in the previous section, as purely for overall guidance. In some courses in modelling, for example in the Open University Mathematics foundation course,[7] a flow chart is used to describe a framework and students are encouraged to fill in the empty boxes in a blank copy of the chart. The majority of participants in the workshop were not greatly in favour of this approach, as experience had shown that there is a fine dividing line between a framework and a straitjacket. Many students tend to see the framework as a 'modelling algorithm' and fail to develop the initiative and flair which is necessary in modelling. At least one of the participants[8] prefers to use a much simpler diagram in explaining the underlying themes:

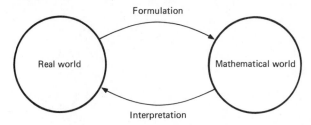

1.6 SOME FINAL ADVICE TO GIVE TO THE STUDENT

Do not expect every problem to fit neatly into the nine stages described in the previous section. Furthermore, do not expect to go through them one by one as you would go through the steps of solving an equation. Backtracking, adjustment and repetition are all quite normal!

Do not rush too quickly to the computer to beat the problem into submission. Even when a computer is necessary to obtain quantitative results it may still be possible to obtain qualitative results by pencil-and-paper methods.

If a computer is to be used, plan your work in advance and try to anticipate the results to be expected.

Do use your initiative and be prepared to discuss all stages of your work with colleagues and teachers.

REFERENCES

1. *The Concise Oxford Dictionary of Current English* (Oxford, Oxford University Press, 1976) p. 701

2. *Webster's Third New International Dictionary* (Chicago, Encyclopaedia Britannica Inc., 1971) p. 1451

3. *Collins' Dictionary of the English Language* (London and Glasgow, Collins, 1979) p. 947

4. R.R. McLone, *The Training of Mathematicians* (London, SSRC Report, 1973)

5. Sir James Lighthill Presidential address, *2nd International Congress on Mathematical Education*, ed. A.G. Howson (London, Cambridge University Press, 1973) p. 88

6. B. Clarke, *The Pursuit of Stillwater Trout* (London, Adam & Charles Black, 1975) p. 15

7. *Open University Course M101: Block V, 'Mathematical Modelling'* (Milton Keynes, Open University Press, 1978)

8. D. Burghes, Mathematical Modelling: A positive direction for the teaching of applications of mathematics at school, *Educational Studies in Mathematics*, 11, 113-131 (1980)

2.

Crop Yield and Number of Plants

A. Barnes

2.1 INTRODUCTORY BACKGROUND

Every time a farmer, vegetable grower or fruit grower, or even a gardener, decides to grow a crop he must decide how many plants are to be planted on a given area of ground surface. Too few plants will mean that, although the weight of crop from each plant may be large, the yield from a given area ground surface may not be very high. Too high a plant population density can lead to weak and spindly plants and the parts of value, e.g. fruits, roots, grain, etc. will be reduced in size, and possibly in number and quality as well. A certain reduction in the value of the crop from each plant can be tolerated because of the compensating effects on yield per unit area of high plant numbers. However, at very high plant numbers yield per unit area can be reduced because of the adverse consequences of small and weak plants. Excessively large plant numbers also tend to increase the chances of disease and pest problems and can generally increase management problems and costs.

Very often the choice of the optimal number of plants per unit area of ground surface does not present a difficult problem for the farmer or grower. If a well tried and trusted crop species is grown, the optimal planting density may be well documented, although examples exist where detailed experimental investigations have shown that substantial yield increases can be obtained by modifying conventionally accepted planting densities. If a crop is being introduced to a region for the first time, or if a new variety of a species is developed or cultural practices are changed, the optimal planting density may not be known. Thus, there can be very important practical reasons for investigating the relationship between crop yield and planting density.

2.2 FURTHER BACKGROUND AND STATEMENT OF THE PROBLEM

Besides the direct practical needs to investigate the relationship between crop yield and number of plants per unit area of ground surface, occasions often arise in agricultural research when it is necessary to determine the biological basis for the relationship between crop yield and planting density. Because of the biological complexity of the plant (e.g. fruit, grain, roots, etc.) it is simplest to consider the biological basis for this relationship by taking yield to mean all parts of all plants within a crop rather than just the economically valuable components.

Further, the way in which the plants within a crop increase in size from the time of planting to harvest is the fundamental biological process which determines the relationship between crop yield and planting density. After planting, or after seedling emergence if seed are sown, growth of the individual small plants is not influenced by the presence of neighbouring plants and the rate at which their weights increase is approximately proportional to their weights. However, as plants increase in size, the leaves of neighbouring plants begin to overlap; the time at which this process begins and the extent of leaf overlap which occurs will depend upon how many plants are grown per unit area of ground surface. The overlapping of leaves implies partial shading of some plants by others. Except for cases of extremely low plant densities, the whole of the planted area will be covered eventually by leaves and all plants in the crop will be shaded to a greater or lesser extent.

This process of ever increasing overlap of leaves and partial shading of neighbouring plants reduces the average intensity of light falling on leaves and leads to reduced levels of photosynthesis, and consequently to reduced plant growth. This process is commonly referred to as plants 'competing' with each other for light. Ultimately, 'competition' for light becomes so intense that the level of photosynthetic activity is only sufficient to replace losses due to leaves dying, etc. and growth ceases completely. An analogous process to competition for light can operate below ground with roots of different plants competing for water and soil nutrients.

The modelling problem is to construct a model to describe the yield (total plant material) of a crop at any time during the period from planting to harvest and to include in this model the effect on yield of differing numbers of plants per unit area. The model should, as far as possible, depend on parameters or terms which have some biological interpretation. Since the main purpose of the model would be to derive from experimental data estimates of these biological parameters or terms which are important in determining the intensity of competition, the smaller the number of parameters or terms consistent with describing the processes involved the better.

2.3 NOTES TO LECTURER

2.3.1 INTRODUCTORY BACKGROUND

Although the details of the modelling problem have not been spelt out in the Introductory Background, it is possible that the lecturer may wish to present that short and rather vague section to students and ask for their ideas about finding the optimal planting density for a new crop. The Introductory Background could be presented with or without the small set of data given in Table 2.1.

Table 2.1 Relationship between planting density and yield

Planting density (No. plants per square metre)	Yield (kg/square metre)
5	2.40
10	4.70
20	9.26
30	13.53
40	16.73
50	18.84
70	16.83
100	15.06
150	13.10

Students might respond to this by suggesting that the variable costs of growing the crop (including the probability of disease and pest problems) need to be quantified. (As an approximation these costs might be assumed to be directly proportional to plant numbers.) Growing costs could then be related to the yield/planting-density relationship, either by considering this graphically or by finding an algebraic expression for yield as a fraction of plant numbers. In either case the importance of the yield/planting-density relationship will become apparent.

2.3.2 FURTHER BACKGROUND

Table 2.2 gives a set of experimental data for barley taken from Aspinall and Milthorpe.[3] The data are the yields per plot taken at each of five times of harvest for crops grown at each of six planting densities.

Table 2.2   Yields per plot for barley

		Number of plants per plot					
		4	8	16	32	64	128
	2	0.8	1.8	2.9	5.8	10.9	15.4
Weeks	3	2.7	6.6	9.8	14.4	17.3	25.6
from	4	8.9	11.9	19.5	26.2	32.0	38.4
sowing	5	14.7	21.8	35.5	39.0	42.9	52.5
	6	17.9	24.0	43.5	52.8	47.4	57.6

If it is thought that students require more information on the pattern of crop growth than is contained in the Further Background the above data set could be supplied. Note that the data above include experimental error and it would not be expected that any derived model would exactly describe the yields. It may assist students when deriving models to be reminded that in almost all areas of biological growth the concept of relative or specific rates of increase of size or numbers is more fundamental than absolute growth rates.

2.4   POSSIBLE APPROACHES TO DEVELOPING MODELS

If students are given the experimental barley data of Table 2.2, then the initial approach may well be graphical inspection of these data. If this is done it will be seen that the relationship between yield and number of plants is initially linear but that as time progresses the rate of increase of yield as plant numbers increase is greater at low plant numbers. Eventually yield is asymptotically related to the number of plants per plot.

It is possible that the next stage in tackling the problem will be to search for an algebraic function which describes families of curves similar to the observed relationships between yield and plant population density. If the search for an algebraic function is not based on any biological justifications, it is probable that the parameters of the function will not have any clear biological interpretation. The model would not then completely fulfil the requirements.

An alternative approach is to consider the rate at which plants in the crop grow at different times and consider how this growth rate may be influenced by the number of plants in the population. If an equation relating plant growth rate and number of plants can be constructed, it can be integrated from time of planting to any time between planting and final harvest. The resulting equation can then be converted into an equation relating the three variables: yield, number of plants and time between planting and harvest.

2.5 DERIVATION OF A MODEL

Two important properties of plant growth within a crop are brought out in the background information:

(a)   the growth of non-overlapping plants;

(b)   the existence of a maximum crop yield at which net growth of plants within the crop ceases.

The model is constructed by considering these two points in greater detail.

Let d be the number of plants per unit area of ground surface, t, the time ($t = 0$ at time of planting), y the yield of crop per unit area of ground, and w the average plant weight (i.e. $w = y/d$).

In almost all cases of biological growth the relative (or specific) rate of growth is more fundamental than the absolute growth rate. It is, therefore, probably more useful to try to construct an equation for relative growth rate $(1/y)(dy/dt)$ of a crop rather than the absolute growth rate dy/dt.

At very low numbers (when d is near zero and thus y is also negligible) plants do not interfere with each other and as an approximation the rate of growth of a plant may be written as

$$\frac{dw}{dt} = kw \qquad\qquad [2.1]$$

where k is a constant. Since $y = wd$ Equation [2.1] can be written as

$$\frac{1}{y}\frac{dy}{dt} = k$$

which applies when the plant numbers are low.

When plant numbers are high (i.e. when d is large) the degree of overlapping of leaves in a crop can be so intense that growth ceases completely. The amount of leaf overlap is likely to be largely dependent upon the amount of foliage in the crop, which is closely related to the total yield. Thus, again as an approximation but nevertheless probably providing a reasonable description, we may write

$$\frac{dy}{dt} \to 0 \quad \text{as} \quad y \to Y$$

where Y represents a 'ceiling' or maximum yield.

If $dy/dt \to 0$ as $y \to Y$ then $(1/y)(dy/dt)$ will also tend to zero as $y \to Y$. Thus as y goes from 0 to Y, $(1/y)(dy/dt)$, goes from k

to 0. No information is supplied (and, indeed, little is accurately known) about what is likely to happen to $(1/y)(dy/dt)$ at values of y between 0 and Y. Following the maxim that when it is necessary to make arbitrary assumptions the simplest ones should be tried first, it is reasonable to assume that $(1/y)(dy/dt)$ declines linearly from k to 0 as y increases from 0 to Y. This leads to

$$\frac{1}{y}\frac{dy}{dt} = k\left(1 - \frac{y}{Y}\right) \qquad [2.2]$$

which can be integrated to give

$$y = Y\{1 + [(Y - w_0 d)/w_0 d]\,e^{-kt}\}^{-1} \qquad [2.3]$$

where w_0 is the value of w at $t = 0$.

For any practical situation $w_0 d \ll Y$, so Equation [2.3] can reasonably be replaced by

$$y = Y[1 + Y/(dw_0 e^{kt})]^{-1}. \qquad [2.4]$$

Equation [2.4] provides a model for yield y per unit area in terms of the two variables, plant population density d and time from planting t. It depends on three parameters Y, w_0 and k, and it can be rearranged to clarify the model. Equation [2.1] gives the assumed form of the growth rate of plants which are growing in the absence of competition from neighbouring plants. Let u be the weight of such a plant. Then

$$\frac{du}{dt} = ku$$

giving

$$u = w_0 e^{kt} \qquad [2.5]$$

as an equation for the weight of an isolated plant at any time.

Substituting Equation [2.5] into Equation [2.4] gives

$$y = Y[1 + Y/(ud)]^{-1} \qquad [2.6]$$

and rewriting this equation in terms of w $(= y/d)$ and rearranging gives

$$\frac{1}{w} = \frac{1}{u} + \frac{d}{Y}. \qquad [2.7]$$

Equation [2.7] provides a simple algebraic function as a model for w in terms of d and t.

More general derivations of Equation [2.7] can be made. For example, if it is thought unreasonable to assume that k, in Equations [2.1] and [2.5], is constant, then k can be any function of time and kt in all previous equations becomes $\int k\,\mathrm{d}t$. Thus, if periods of cold, dry or generally unfavourable growing conditions occur so that isolated plant growth is not completely exponential, as must be expected in practice, the model provided by Equation [2.7] may still be applicable, although the exact relationship between u and t may not be known.

2.6   INTERPRETATION OF THE MODEL

The final form of the derived model, Equation [2.7], states that the reciprocal of the average plant weight w is a linear function of plant population density d. The slope of this relationship is the reciprocal of the ceiling yield parameter Y and the intercept is the reciprocal of the weight of an isolated plant u. Although the model relates plant weight (or yield per unit area) to both plant population density and time from planting, the latter variable, represented by t, does not explicitly occur in the model. The complete effect of time from planting on the yield/population-density relationship is contained in the term u, representing the isolated plant weight at time t.

Estimation of the parameters in the model from experimental data should ideally be done by statistical curve fitting procedures after careful consideration of the likely pattern and magnitude of experimental error. However, approximate estimates can be arrived at by plotting the reciprocal of average plant weights against planting population density (see Fig. 2.1).

Assuming linear relationships between $1/w$ and d with a common slope at each harvest time as implied by the model [2.7], the parameter Y is estimated to be 51.2. The way in which u varies with time of harvest could be investigated by considering the intercepts of the linear $1/w$ versus d relationships. By visual inspection it is clear that u does not increase exactly exponentially, although discrepancies are not very large.

The reciprocal plot of the data suggests that there are systematic differences between the model and the observed data. This is hardly surprising in view of the simplicity of the derived model. Discrepancies of the order of magnitude seen in this plot are sufficiently small for the model to be of practical value.

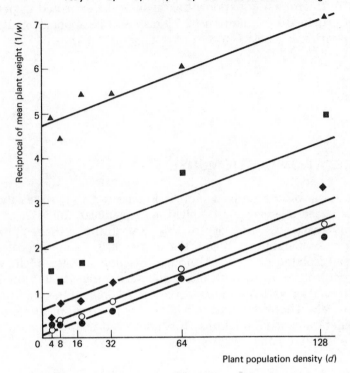

Fig. 2.1 Barley data from Aspinall and Milthorpe (1959)

Key: ▲ 2, ■ 3, ◆ 4, ○ 5, ● 6 weeks after sowing.

A further set of data, collected from a glasshouse pot experiment on lettuce, is given in Table 2.3.

Table 2.3   Lettuce yield (weight per unit area)

		Plant population density			
		1	9	120	400
Number	18	0.2	2.0	21	50
of	28	2.0	17	53	80
days	38	11	42	83	85
from	48	33	86	105	88
sowing					

2.7   FURTHER DEVELOPMENTS

1. For many crops, especially those grown for human consumption, only a proportion of each plant is useful. The magnitude of this portion

varies with the size of plants. For carrots an experiment gave the following average weights for the useful (root) and useless (leaves) portions of the plants for plants of different sizes.

Average weight of roots (grams)	Average weight of leaves (grams)
3.5	3.9
11.7	9.4
19.6	14.3
34.6	19.0
90.7	35.7
160.6	57.0
297.3	87.8
400.6	111.1

Suggest a possible model for the relationship between the yield of carrot roots and planting density.

2. In practice plants are rarely grown at equal distances from all neighbouring plants. It is much more common for seeds to be sown in rows so that the distances between plants within a row is much less than that between rows. Consequently, leaves of plants growing in the same row will overlap long before leaves of plants growing in different rows. Plants are capable of adapting their physical shapes, to some extent, to make the best use of the available ground surface, but their ability to do this varies considerably from species to species.

Suggest ways of modifying the model(s) for describing the relationship between yield and planting density to take account of differences in patterns of planting as well as differences in density.

3. The whole process of plant competition becomes more complex and difficult to handle mathematically when competition between species is considered, e.g. crop and weed competition, mixtures of crop species or healthy and diseased plants. Suggest possible ways to approach this aspect of competition.

REFERENCES

1. R.W. Willey and S.B. Heath 'The Quantitative Relationships between Plant Population and Crop Yield', *Advances in Agronomy* 21, 281–321 (New York, Academic Press, 1969)
   This paper gives a good general review of the mathematical aspects of plant population–crop yield relationships and a large number of references to experimental and theoretical work in this area.

2. A. Barnes 'The influence of the length of the growth period and planting density on total crop yield', *Annals of Botany* **41**, 883–895 (1977).
   This paper gives derivations to a number of yield — population density — time models along similar lines to the derivation presented here.

3. D. Aspinall and F.L. Milthorpe 'An analysis of competition between barley and white percicaria', *Annals of Applied Biology* **47**(1) 156–172 (1959).
   This paper provides useful data and an interesting way of looking at competition between different species of plants.

4. M.A. Scaife and D. Jones 'The relationship between crop yield (or mean plant weight) of lettuce and plant density, length of growth period, and initial plant weight,' *Journal of Agricultural Science, Cambridge* **86**, 83–91 (1976)
   This paper provides useful data and discussion.

3.

Deer Harvesting

J.J. McDonald

3.1   GENERAL BACKGROUND AND MOTIVATION

The following information is provided as background and motivation for teacher and student.

In recent years it has become widely recognised that the Earth's resources need to be carefully managed. There are two broad categories of natural resources — renewable and non-renewable. Non-renewable resources include things such as fossil fuels (e.g. coal and oil) and metal ores. Renewable resources are those which are capable of regeneration such as trees, crops and animal food sources.

Reasonably reliable estimates can be made of the available amounts of each non-renewable resource, and attempts at managing these resources have included efforts to damp down demand, reclamation and substitution by renewable resources. Although renewable resources have the advantage that they are capable of regeneration, the time scale of regeneration is often long in comparison with that of exploitation and, if over-exploited, the resource may be lost forever. A well publicised example of this is the over-exploitation of whales. There are many factors affecting the regeneration of a renewable resource and the problem of planned exploitation becomes a very subtle one.

The failure of the American wheat crop several years ago with its world-wide repercussions on food prices is an example of the susceptibility of a natural resource to disease. Planning is therefore difficult but questions on how to exploit resources arise continually.

At the time of writing, a controversy exists on the policy to be adopted during the mackerel season off the coast of Scotland. The instinctive reaction of the fishermen — based partly on memories of the disappearance of the West Coast herring shoals — is that fishing should be done on only five days per week. The argument of the fleet owners in favour of a seven day fishing week, is economic and supported, it is claimed, by scientific evidence that this will not be to the detriment of the resource. Who is correct? Other topical examples are never difficult to find. For example, the restriction of the 1979 grouse shooting season because of the shortage of birds following the severe winter conditions early in the year and the effect of a disease carried by the sheep tick.

Can you think of other examples?

Whilst some kind of management is clearly desirable, the effect of certain strategies can be counter-intuitive. One of the best documented cases concerns the deer population of the Kaibab plateau in Arizona.[1] In 1907 a bounty was placed on the natural predators of these animals. (Trappers were paid a fee for each predator skin returned.) The effect of this was that the predator populations fell rapidly, and by the early 1920s the deer populations had risen to a very high level. A consequence of this was that their grazing areas became overbrowsed. Despite growing evidence and calls for action, nothing was done to control the deer population and in two severe winters the population was drastically reduced. That was some time ago, but in the 1976 Annual Report of the Red Deer Commission[5] there was a plea to estate owners to reduce the stocks of deer on their estates to avoid the heavy mortality and damage to agriculture and forests which would inevitably follow a hard winter.

3.2 STATEMENT OF THE PROBLEM

Examine how mathematical modelling methods might be applied to the development of harvesting policies for the control of deer herds in woodland areas where the primary objective of management is the production of good quality timber.

The following points should be noted.

(a) Give close attention to the formulation of the problem.

For example, what is the objective of the model, where will you find background information, what data is available and what has to be collected, what are the fundamental relationships?

(b) Keep the model simple.

(c) When you have finished the model and analysed it, assess the contribution it has made to your understanding of the problem.

(d) State the limitations of the model and discuss whether you think it would be worthwhile to develop a more complicated model. If so, what additional factors would it include, what would be its purpose and would it be radically different from the first?

3.3 GENERAL COMMENT TO THE LECTURER

The problem statement is deliberately vague and is presented in a way in which mathematicians are often given problems. Thus the study could form a substantial piece of project work for advanced students. As we proceed through the study, however, it will become evident that there are sub-problems which are well within the capabilities of more junior students.

At the present stage students should be given considerable time to think about the problem and acquire the basic information which is required to proceed. The lecturer may like to play the part of the 'client' — for example one of the woodland management team or a biologist. This approach gives the students opportunity to develop skills in eliciting information through discussion with a client.

References 1, 2 and 4 will be of use at this stage as a source of background information. Reference 13 discusses aspects of communication between mathematician and client and should add some overall perspective to the study.

3.4 MODEL DEVELOPMENT

Before we can even think of building the model, there is much to be done. We have to learn more about the problem, decide on the aspects of the problem with which our model is going to deal and then formulate the problem in a form in which it can fairly readily be translated into mathematical terms.

3.4.1 DISCUSSION – THE FIRST STEP IN UNDERSTANDING AND FORMULATION

One of the first steps in the modelling of a new situation is often a discussion between mathematician and client, whether the client be a commercial client or research associate. At this stage the mathematician starts to develop an understanding of the real system and takes the first steps in the formulation of the problem.

The following summary is typical of the outcome of such discussions. It is often useful in such dialogues to reach an agreed written statement of the situation. The summary is based on actual discussions between the author and a biologist working in this field.

> The discussions centred around the management of red deer in forest habitats. The main commercial interest of the woodland management is in the production and marketing of trees. Forested areas can provide a high-quality habitat for red deer which frequently respond with high reproductive rates. In the absence of predators, other than man, populations can increase rapidly,

leading to damage to valuable tree stocks. The management therefore has a policy of controlling deer populations. Although this is done primarily to protect trees the harvested deer have a significant commercial value. The management is conscious of its impact on the environment and therefore wishes to carry out an informed culling programme.

Culling is carried out by stalkers who have to be advised on the numbers of deer to be removed from each area during each season. One aim of current research is to produce a model which will predict the effect on the population of various culling regimes.

To predict the effect of culling on a deer herd the population dynamics have to be studied. At its simplest level, deer are born, move through various age classes with the possibility of dying naturally or being shot at any stage in this process. The qualitative effect of culling policies is easily demonstrated but the production of reliable quantitative results from a model depends upon the ability to obtain good estimates of population levels, immigration and emigration, recruitment and mortality rates. All are difficult to obtain but a degree of confidence is accumulating with respect to values available as a result of current research work.

Absolute numbers of deer are extremely difficult to obtain and in woodlands must be estimated by indirect methods, i.e. pellet counts, drive counts, etc. Approximate figures are available for some forest areas.

It is impossible to categorise a herd into year classes, though it is possible to distinguish calves, yearlings and adults of each sex which have died naturally or been culled following the autumn rut. Fecundities are obtained from postmortem examinations of females and are variable from area to area. In some forests most female yearlings are pregnant whilst in other forests pregnant yearlings are never encountered. Precise fecundity rates for most age classes are available for most areas; these data constitute the most reliable available for modelling.

The quality of the habitat is an important determinant of the size of the population which a forest will maintain. Very young plantations offer food without cover, pre-thicket offers food and cover while thickets and high forest offer cover but little food.

An adequate amount of cover and food is important but high densities can only be supported where plantation and pre-thicket predominate. Most of the commercial forests in Scotland are comparatively recent and have not reached a steady state in their evolution. Forecasts of the structure of the forests in the future suggest that they will be dominated by high forest rather than as at present by pre-thicket and thicket. It is possible, that as the structure of the forest changes there will be less vegetation to support the deer. One possible consequence of this is that the deer may turn to the trees as a source of food.

3.4.2 OBJECTIVES OF THIS MODELLING EFFORT

From reading the previous section and the appropriate references, it should become clear that a model is required which will describe the changes in deer population in reaction to natural effects and management policies. Criteria have to be developed for use in obtaining better management policies.

It has been advocated by others[14] that the ultimate form of the model should only be considered late in the formulation stage, lest the methods to be used in the analysis of the model over-influence the formulation of the problem. In what follows, we will attempt to follow the spirit of that approach.

3.4.3 FORMULATION

In modelling exercises of this type we are dealing with a realistic rather than a real situation. Students will not normally have the opportunity of discussing the problem with a real client. Often, the best that can be done is for the student to discuss the problem with a lecturer who plays the role of client. Inevitably this will lead to different formulations of the problem depending on the assumptions made early on. The following formulation is therefore not definitive but will allow us to continue.

The main objective of woodland management is to produce good quality timber and maintain commercial viability in the process. It is argued by some experts that deer damage the timber and thus are detrimental to the commercial value of woodland areas. One alternative open to wood-land management would be to eradicate deer herds from woodlands. For general conservation reasons, however, this alternative is not politically acceptable. We have seen earlier that the management 'wishes to carry out an informed culling programme'. This presumably means that they want to maintain the deer population at an acceptable level — however defined — and need to know the relationship between population structure and culling policy. This would be of particular value if the management decided to examine the viability of commercial exploitation of the deer.

We will therefore start by looking at the relationship between culling policies and population structure.

Some time spent at this stage on the discussion of what is meant by 'culling policies' will be of value.

We saw earlier that culling is carried out by stalkers 'who have to be advised on the number of deer to be removed each season'. Clearly instructions have to be given by management, based on some pre-planned strategy. Not only have the stalkers to be told how many deer in total are to be removed, but also the sex and approximate age of the individual animals to be killed. At the simplest level, therefore, a culling policy is a planned removal of deer in each age and sex group. A model which represents how the population structure changes in reaction to different culling decisions would be of great value.

The next step in formulation of the problem is to examine the 'dynamics' of the population (i.e. how the structure changes with time) in the absence of culling. The following simple statements provide a starting point.

- The population may be stratified into sex and age classes — although it may be difficult for a stalker to identify, at a distance, the age class to which a particular animal belongs.

- A proportion of hinds (females) in each age class (other than calves) will produce offspring.

- Each hind produces a single calf.

- Calves are born during a period of six weeks in June and July following an autumn rut.

- Each year a proportion of animals in each age group will die from natural causes.

The key variables therefore are the following:

- the number of animals in each age class in any year;

- the proportion of hinds in each age class producing offspring each year;

- the proportion of animals dying each year in each age class.

In many cases it is considered that the various proportions referred to above are constant from year to year for any given class. In other cases it is argued that the proportions depend on factors like weather and population density.[4]

3.5 MODEL 1

3.5.1 FORMULATION

A few fairly non-restrictive assumptions will now allow us to develop a simple model describing the population dynamics. We will assume that the ratio of males to females is the same in each age group and remains constant throughout the history of the population and that the birth and death proportions referred to in the formulation are the same for both sexes at all ages.

Assume that there are $m + 1$ age groups as follows:

Group 0 = Calves

Group 1 = Yearlings (age 1 last birthday)

Group 2 = Hinds aged 2 last birthday

Group m = Hinds aged m last birthday.

Note that we consider only the hind population. Under the above assumptions the stag population can be inferred from the hind population.

Let the proportion of hinds in age group x at time t surviving to age group $x+1$ at time $t+1$ be P_x, where $P_x > 0$ for $x < m$ and $P_m = 0$.

Let F_x denote the proportion of hinds in age group x at time t giving birth to a female calf.

Let the number of hinds in each age group at time t be $n_0(t), n_1(t), \ldots, n_m(t)$.

Then if we consider the start of year $t+1$ to coincide with the start of the calving season, the number of female calves in year $(t+1)$ is

$$F_1 n_1(t) + F_2 n_2(t) + \ldots + F_m n_m(t) \equiv n_0(t+1).$$

The number of hinds in each age group $i > 1$ will be

$$P_{i-1} n_{i-1}(t) \equiv n_i(t+1), \quad i = 1, 2, \ldots, m.$$

We thus have a model for predicting the population structure in year $(t+1)$ given its structure in year t.

If we consider the population structure in year t to be described by the $(m+1)$-dimensional vector $n(t)$, then the structure in year $(t+1)$ is given by $n(t+1)$ where

$$n(t+1) = Mn(t)$$

$$\text{and } M = \begin{bmatrix} 0 & F_1 & F_2 \ldots F_{m-1} & F_m \\ P_0 & & & \\ & P_1 & 0 & \\ & & P_2 \cdot \cdot_{\cdot} & \\ 0 & & P_{m-1} & 0 \end{bmatrix}$$

Thus if $t = 0$ is used, arbitrarily, to denote the start of a period of study, the population is described at some future time $t = T$ by

$$n(T) = M^T n(0)$$

The latter form of the model is more amenable to analytical treatment than the first form which is more suited to computational work.

We now have a simple model which we can use to examine the behaviour of deer populations. We have seen already in the section entitled 'summary of discussions' how total populations are estimated in forest environments. Readers interested in the details of how birth and death rates can be obtained are referred to the paper by Lowe[4] which records the results of a long term study carried out on the island of Rhum. Although most forest environments differ from that on Rhum it should be clear that, if enough care is taken in data collection, this simple model can have a high degree of realism.

The data of Tables 3.1 and 3.2 (which are hypothetical but typical of those found in practice) represent the situation at a particular time in a forest area of about 2000 hectares   (1 hectare $= 10\,000\,\mathrm{m}^2$).

Table 3.1   Deer population structure

Stag count	Hind count	Age last birthday
48	47	0 (Calves)
32	33	1 (Yearlings)
23	22	2
15	15	3
15	15	4
8	9	5
3	3	6
3	3	7
1	2	8
2	1	9
1	1	10
0	0	11
0	0	12
0	0	13
0	0	14
0	0	15

A hand calculation or simple computer program, using the data of Tables 3.1 and 3.2 and following the previous theory, produces the population structure given in Table 3.3. This table shows the population after five years and again after ten years from the start of the simulation, the start corresponding to year zero.

Table 3.2 Birth and death data

Year class	Proportion of hinds dying	Proportion of hinds calving
0 (Calves)	0.15	0
1 (Yearlings)	0.10	0
2	0.05	0.75
3	0.01	0.95
4	0.01	0.95
5	0.01	0.95
6	0.01	0.95
7	0.01	0.95
8	0.02	0.85
9	0.05	0.75
10	0.05	0.70
11	0.15	0.50
12	0.20	0.20
13	0.50	0.10
14	0.80	0.10
15	0.90	0.10

Proportion of female calves born is 0.5

Table 3.3 Simulated population structure

Year class	No. of hinds		
	Year 5	Year 10	(Now ≡ Year 0)
0	80.0	186.1	
1	58.1	134.1	
2	44.5	102.9	
3	31.5	83.1	
4	22.4	69.0	
5	33.5	57.0	
6	27.4	48.2	
7	20.1	40.6	
8	14.3	30.0	
9	14.1	21.1	
10	8.1	30.2	
11	2.6	23.7	
12	2.2	14.9	
13	1.2	8.6	
14	0.3	4.3	
15	0.1	0.5	

The above data and calculations illustrate how the model may be used to simulate the dynamically changing population structure.

This kind of model occurs frequently in the study of biological populations. The matrix M in the matrix form of the model is the so-called Leslie Matrix.

The theoretical properties of this model will not be considered further here. A treatment of these may be found in Pollard.[11] From a practical point of view, however, the most interesting result is that the population vector $n(t)$ eventually takes on the structure of the dominant eigenvector of M.

The population dynamics are then governed in the long term by

$$n(s+1) = \lambda n(s), \quad \text{for some } s,$$

where λ is the dominant eigenvalue of M.

Such an age distribution is *stable* in the sense that the ratio of the number in one age class to that in another remains constant from one year to the next, and $n(s)$ is 'the vector of the stable age distribution'.

The population is stationary if $\lambda = 1$, growing if $\lambda > 1$ and falling if $\lambda < 1$, whilst still maintaining its stable age distribution.

3.5.3 MODEL VALIDATION

A significant step in the development of a model is its validation. This is the process of checking that the model does adequately represent the system being studied. In the present situation we would, ideally, like to check that the model is capable of predicting the deer population structure. Unfortunately, there are only two ways of doing this. One is to predict the population for a number of years in the future — probably for at least as long as the maximum lifetime of a deer — and then carry out counts of the herd in each one of these years. The other is to make use of historical data and check whether or not the model is capable of replicating the development of the population over the time to which the historical data relates. In both methods we would have to be certain that the conditions used in the formulation and development of the model pertained throughout.

In situations like this, more often than not, neither of these approaches is really practical. Faced with the pressure of making decisions, it is seldom the case that management can afford the luxury of the long-term experimentation required in the first approach. In the second approach — using historical data — we often find that there are few data available or that insufficient care has been taken in collecting the data or that conditions have changed during the period for which the data are available.

There are also the practical difficulties of carrying out accurate population censuses and estimating birth and mortality factors. Lowe,[4] for example describes in some detail the attempts which were made at obtaining a reliable picture of the deer population on Rhum. Briefly, these involved carrying out several censuses between 1957 and 1966 and a careful accounting for each deer which died during this period in an attempt to account for all of the original 1957 population. By 1966 almost 92% of the 1957 population had been accounted for and the data obtained were used by Lowe to produce life tables.

It should be pointed out, however, that a programme of culling was carried out during the period of the Rhum study so the data are not relevant to the basic model described in this study.

Even in situations where data are not available for validation of the model, the model can be useful. Assuming that the model does reflect the mechanism of movement of the population, uncertainty is likely to centre around the census of the original population, the proportion of hinds giving birth and the proportion surviving from one year to the next. The model, however, can be used to examine the sensitivity of the population structure to variations in the data. This is particularly important in developing a new culling policy in the absence of reliable data. In the early stages, it might be useful to know whether the cull of any age and sex group should be of the order of, say, 5% or 25%. It should be accepted therefore that it may take years for the model and the associated data to evolve from the state where they give only qualitative information, to a stage where reliable quantitative information can be given.

3.5.4 FURTHER DEVELOPMENT OF MODEL 1

In the light of what has been said above about validation, the student at this stage might be encouraged to do some experimentation on the basic model with a view to developing a better understanding of its behaviour. The final stage is to examine how the model can be modified so that it can contribute to the formulation of culling policies. Such developments can be done by the student in the form of exercises, and a few examples are given below.

The remainder of the study, however, goes on to look at another model structure which could arise from the same formulation. The main lesson to be learned here is that, on completion of the formulation, a unique model is not necessarily obtained. In practice the final structure of the model will depend greatly on the model builder's experience and patterns of thought. The reader should compare the arguments leading to each model and assess the contribution of the model to the problem in hand. In the end, both models give identical results. The difference lies in the way in which the dynamics of the population are described.

Some readers may find model 2 rather more difficult to follow or to implement on a computer. These readers may wish to skip model 2 and concentrate their efforts on the questions which follow on model 1. Appendix 3.1 gives some hints on handling these questions.

3.5.5 FURTHER QUESTIONS ON MODEL 1

1. Given that deer have been observed in some number in this forest over a period of 30 years and that little shooting has taken place, does the behaviour of the model seem plausible?

2. Some biologists suggest that, for some habitats, as the density of deer increases so the death rates increase and the birth rates fall.[4] Modify the basic model to take account of this feature of the situation. Hypothesise some relationships between birth and death rates and density and examine the development of the population under these hypotheses.

3. Returning to the basic model, assume that the model and its associated data are valid. Discuss what is meant by a 'deer management policy'. Modify the basic model to reflect specific management policies and examine the behaviour of the population under these policies.

4. To what extent would management policies have to change if density effects were present?

3.6 MODEL 2

3.6.1 FORMULATION

In developing this model we start with the same formulation and data requirements as for Model 1. This time we will attempt to create a picture of the situation as a whole by keeping track of where everything goes and explicitly recognising the dynamic nature of the system with which we are dealing. In doing this, we will view the number of animals in each class as representing the state of the system in any year. This state will change through flows of animals into or out of a class and from one class to another. The description of the situation (or system) which we will develop will, in the first instance, be visual rather than mathematical.

The author is quite certain that this kind of approach is unlikely to be the first to occur to the student — or, indeed mathematician — who has not used it before. That used in developing Model 1 is likely to arise much more naturally. Students of mathematical modelling should, however, at some stage, be encouraged to create total pictures of systems because they (the students) tend to be weak in overall perspective. There are also practical advantages to non-mathematicians in developing visual models of their systems and problems. Unfortunately, the relative ease with which such models can be created has meant that there have been many abuses of this approach. In fact, some readers may feel that, having derived what seems to be an adequate model already, the use of a new approach is not appropriate here. We use it in the following light.

(a) Students, at an appropriate stage, like it, use it well and learn much about the wider issues in modelling.

(b) It can be a powerful and useful approach in complex situations — particularly (as in this case) where data are lacking.

(c) Non-mathematicians derive a lot of practical benefit from it.

We will follow System Dynamics Modelling conventions which have been described by Forrester[7-9], Pugh[12], Coyle[6] and Goodman[10].

To simplify the description of the model, we will consider five classes of deer — calves, male yearlings, female yearlings, mature males and mature females. It will subsequently become clear that a further stratification into year classes is easily carried out.

The model may be represented in the diagrammatic form of Fig. 3.1, where the various symbols have the following meanings.

represents a *level* variable such as the number of mature males at any time.

represents a *rate* such as the death rate (animals/unit time) of mature females.

represents a constant.

Dottled lines – – – – – – – → represent flows of information, and

Solid lines ——————→ represent flows of animals.

The variable names are in the form of mnemonics and have the following meanings.

LEVELS

CAVS   calves (unit is animals)
FYRLG   female yearlings (animals)
MYRLG   male yearlings   (animals)
FMATR   mature females   (animals)
MMATR   mature males   (animals)

RATES

CBR   calf birth rate (unit is animals/year)
FYRR   female yearling recruitment rate   (animals/year)
MYRR   male yearling recruitment rate   (animals/year)
FMRR   recruitment rate of mature females (animals/year)
MMRR   recruitment rate of mature males   (animals/year)
FMDR   death rate of mature females   (animals/year)
MMDR   death rate of mature males   (animals/year)
FYDR   female yearling death rate   (animals/year)
MYDR   male yearling death rate   (animals/year)
CDR   calf death rate   (animals/year)

CONSTANTS

BMF   births per mature female (unit is births/female/year)
FYRF   female yearling recruitment factor   (recruits/calf)
MYRF   male yearling recruitment factor   (recruits/calf)
FMRF   mature female recruitment factor   (recruits/yearling)
MMRF   mature male recruitment factor   (recruits/yearling)
CMF   calf mortality factor   (deaths/calf)
FYMF   female yearling mortality factor   (deaths/female yearling)
MYMF   male yearling mortality factor   (deaths/male yearling)
FMMF   mature female mortality factor   (deaths/mature female)
MMMF   mature male mortality factor   (deaths/mature male)

The sequence of symbols, for example

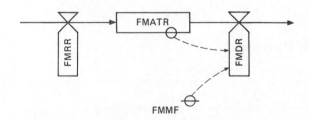

shows that the number of mature females is 'topped up' by yearling recruits and reduced by deaths. The rate of deaths is related to the number of mature females and the quantity FMMF. For example, the death rate might be directly proportional to the number of mature females and FMMT would be the constant of proportionality. The diagram is otherwise self-explanatory and provides a visual model of the system. One advantage of this approach is that the discipline of creating the diagram forces us to consider carefully the structure of the system. It is, however, usually necessary to translate the diagramatic model into

Fig. 3.1 Diagrammatic model for behaviour of deer

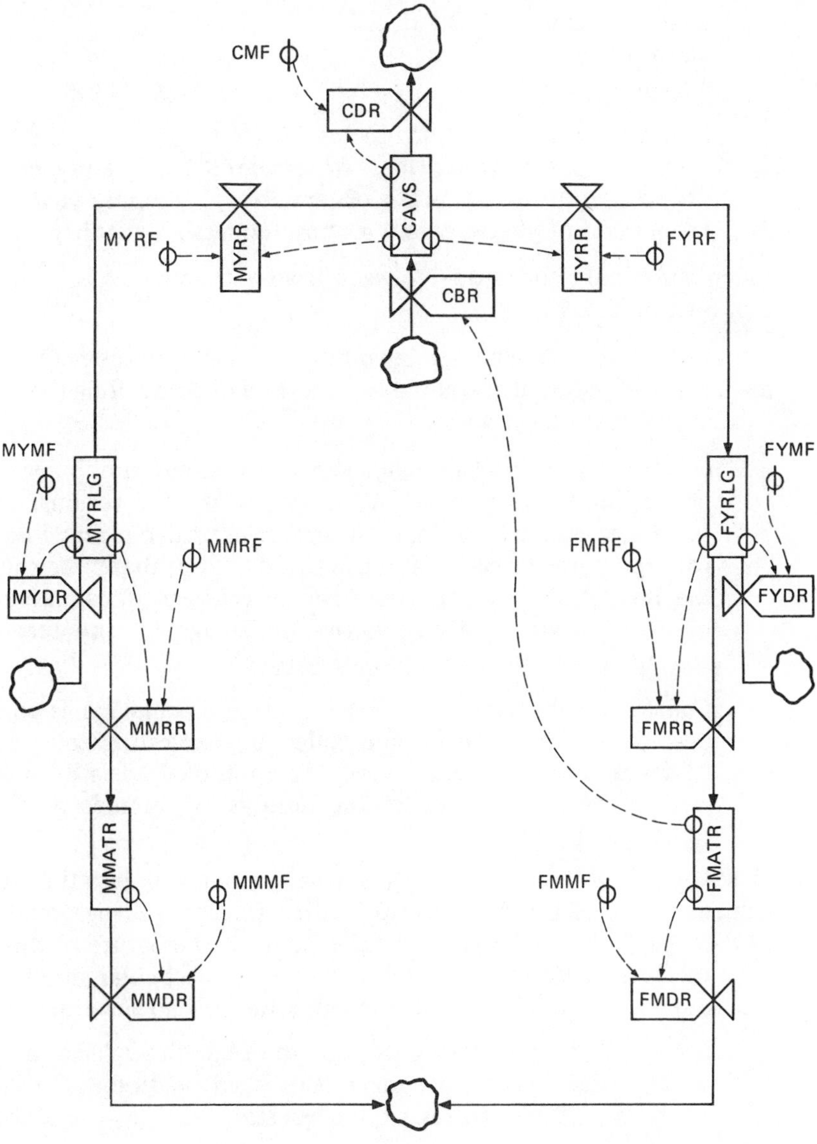

a computer-based model. As in the case of model 1, a program can be written in any common high level language (ALGOL, FORTRAN, BASIC, etc.) but a special langauge, DYNAMO[6,12] has been developed for model structures of this type.

3.6.2 ANALYSIS AND DEVELOPMENT

A very crude aggregation of the data used with model 1 gives

Calves	96
Female yearlings	33
Male yearlings	32
Mature females	71
Mature males	71

Proportion of calves dying in any year	0.15
Proportion of yearlings (male or female) dying in any year	0.10
Proportion of mature males or females dying in any year	0.10
Proportion of yearlings and mature females giving birth	0.70

Using these data and model 2 we will answer some aspects of Question 3 of Section 3.5.5.

The first aspect is what is meant by a 'deer management policy'? A management policy is presumably a method of controlling the population to achieve some objective(s).

Objectives: In a deer management problem the objective might be to limit the population so as to prevent serious damage to forests or agricultural land. Alternatively the herd may be managed for commercial reasons and the aim would probably be to maximise the net return from this commercial venture. Again the aim might be to achieve a balance between damage to one resource (e.g. forests) and return from the other (deer).

Control Mechanism: The basic control mechanism is simple. The deer population can be controlled by harvesting selectively parts of the population. This is done on an annual basis by stalkers who can be given orders as to the number of animals of each group to be killed.

If we consider the first objective — i.e. limit the population to prevent damage — one of the main practical difficulties is relating population size to damage. There is no general agreement on the nature of this relationship. In the absence of such a relationship in a particular situation, a level can be specified which it is felt the environment can sustain.

Specific Policies: Having defined the objective of the management policy, the policy becomes the identification of the control mechanism. This centres around questions such as the following:

Should a fixed number of deer in each group be harvested each year or should a fixed proportion of each group be harvested each year?

Should the proportions of deer harvested in each group be the same?

Should the harvesting be related to the density of animals and if so, how?

The model summarised in Fig. 3.1 needs to be extended to take account of harvesting. A DYNAMO program implementing such an extension appears in Appendix 3.2. This program simulates a policy in which the basic rule is to cull a fixed proportion of each group of animals (Equations [57]-[61]) but this proportion is adjusted upwards or downwards (Equations [21]-[25]) according to whether the density of deer (Equation [17]), as measured by a density ratio, is respectively above or below a specified norm (Equations [20] and [46]). In the case of mature deer, the cull rate is also adjusted to take account of changes in the ratio of stags to hinds (Equations [18], [19], [21] and [22]).

AUXILIARY VARIABLES

TPOP	total population	
DENS	density (animals/hectare)	
RAFM	ratio of mature hinds to stags	
RAMF	ratio of mature stags to hinds	
RDEN	ratio of density to normal density	
FMCF	proportion of mature females culled	(animals)
MMCF	proportion of mature males culled	(animals)
FYCF	proportion of female yearlings culled	(animals)
MYCF	proportion of male yearlings culled	(animals)
CCF	proportion of calves culled	(animals)
FMATC	number of mature females culled	(animals)
MMATC	number of mature males culled	(animals)
FYRC	number of female yearlings culled	(animals)
MYRC	number of male yearlings culled	(animals)
CAVC	number of calves culled	(animals)

Recruitment and mortality factors which appeared as constants in Fig. 3.1 have become auxiliary variables to allow changes caused by culling (Equations [32]-[39]). The corresponding constant figures appear as 'normal' rates, e.g. NFYRF (Equations [48]-[56]). The additional variables have the following significances.

CONSTANTS

NFMCF	normal proportion of mature females culled
NMMCF	normal proportion of mature males culled

NFYCF   normal proportion of female yearlings culled
NMYCF   normal proportion of male yearlings culled
NCCF   normal proportion of calves culled

The program was executed using the above data and a normal cull of 25% of each group of deer (see Equations [40]-[61]. The relevant program output may be seen in Appendix 3.2.

DYNAMO readily provides tabular and plotted output. The simulation represents a management period of 20 years in which total population has been driven down from 303 deer to 87 with a drop in density from 15 deer/100 hectares to 4.3 deer/100 hectares. (Deer/100 hectares is the preferred practical unit for the measurement of density.) The drop in population was rapid initially but was slow towards the end of the period because of the density related harvesting policy. The policy, however, has driven the density well below the level which is considered to be 'normal'.

The qualitative effect of a 12% cull is similar with total population falling to 177 deer (8.9/100 hectares) in 20 years. This gives a density close to the 'normal' density but still with a slowly falling population. As a comparison, a cull of 7% (normal) per annum produces total populations which deviate little from the initial conditions.

Alternative culling policies are easily implemented in DYNAMO, although it should be pointed out that the availability of DYNAMO is in no way essential. DYNAMO is merely a convenient language in which to express the model and with which to obtain output. The model could be expressed in any other high level language but this might be a little more tedious. It is left to the reader to propose and test alternative policies.

3.6.3 FURTHER QUESTIONS ON MODEL 2

1.   Adapt the basic model described in Fig. 3.1 to allow simulation of the movement through year classes (as in model 1) and compare the behaviour of the two models.

2.   Include a harvesting effect in the year class model provided in answer to 1. and repeat the simulations of the worked example.

3.   Modify the basic model (Fig. 3.1) to take account of density-related birth and death rates and examine the behaviour of the population under a variety of hypotheses concerning the relationship between density and birth and death rates.

4.   It is suggested that environmental factors should be taken into account when developing a deer management policy. For example, Lowe[4] states '... the majority of deaths (87.9%) were apparently

due to malnutrition or starvation. These coincided with adverse weather affecting the quantity, quality and availability of food. If we take rainfall as an index to climate, peak periods of rainfall were followed by peaks in mortality...' We also noted earlier that hinds must reach a critical body weight before they can conceive. After giving birth, they lose weight during the nursing of their young and are incapable of conceiving again until they regain condition. The rate at which their condition is regained depends on the availability of food. For example, in ideal forest habitats hinds may seldom fall below critical weight and can conceive each year but on hill estates some hinds may only be capable of conception every second year.

Discuss how these features should be incorporated in the management model — remembering that the reduction of the size of herds by culling may improve the food availability for those left.

3.7 QUESTIONS RELATING TO MODELS 1 AND 2

1. Adapt model 1 to fit the situation of model 2 in which deer are categorised simply as calves, male and female yearlings, male and female mature. Verify that both models have identical behaviour under conditions of (a) no harvesting (b) constant rate harvesting.

2. Examine, analytically, the behaviour of the adaptation of model 1 suggested in Question 1.

3. Discuss how the features described in Question 4 above could be incorporated in a model having the basic features of model 1.

3.8 COMPARISON OF MODELS 1 AND 2

A thorough study of models 1 and 2, including attempts at the questions which have been posed, will reveal advantages of both approaches.

Model 1 is more succinct and mathematically elegant than model 2. It is also easier to adopt an analytical approach to the study of this model. Where simulation is required, it is relatively simple to write a program or indeed carry out the simulation by hand.

Model 2 is much less succinct and more difficult to program (for those unfamiliar with DYNAMO). It has the advantage of providing a clear model building methodology, where the modeller keeps track of what is happening to all of the animals within the system. Once the basic model has been created, it is relatively easy to add new features to the model and the program. (DYNAMO allows the equations to be specified in any order.) Again it should be emphasised that DYNAMO is not essential for the implementation of model 2.

3.9 PRESENTATION OF THE MATERIAL

In all cases work on this material must be preceded by some description of the real system. This can be abstracted from the introductory material to this paper and from the references cited. In this respect Lowe[6] is particularly useful. The background information can be presented in the form of a lecture or reading material. An exercise which is always more demanding for students involves the lecturer in developing suitable background knowledge and playing the client whilst the students attempt to elicit the necessary information. The amount of background knowledge necessary for completion of the task increases with the sophistication and generality of the questions asked.

Modelling assignments of elementary to intermediate difficulty can be organised around the following plan.

(a) development of the Leslie matrix model (model 1);
(b) examining the behaviour of the Leslie model with various sets of data and harvesting strategies.

The same kind of work with system dynamics models is conceptually no more difficult but a new language effectively has to be introduced before progress can be made.

More difficult work involves vague questions of the following type:

What is meant by a management policy?

Hypothesise relationships between A and B and investigate the behaviour of the system under these hypothesised relationships.

APPENDIX 3.1

For those students who may find aspects of Model 2 difficult to follow, here are some hints on how to handle the questions on Model 1.

Question 1 (plausibility of the basic model)

The text gives the simulated population in year 5 and again in year 10. Determine the simulated population structure in year 1, showing that the simulated calf population is approximately 66% of the actual calf population in year 0. Would you expect this kind of behaviour? If not, can you suggest anything that might be wrong?

Note also that the data on survival probabilities and fecundities suggest that the deer can live to around 15 years old. On the other hand, in the population structure given there are no animals over 10 years old. As the text shows, the higher age groups fill up as the simulation proceeds. Since the deer have been present for at least thirty years, does this behaviour of the model now seem plausible?

Re-write the model so that it could be used to estimate what the population structure was 30 years ago. Do this estimation and comment on the results.

Question 2 *(density dependent birth and death rates)*

As in the text, assume that there are $m+1$ age groups. If the area occupied by the herd is A hectares, then in the notation of the text, the density $D(t)$ is given by

$$D(t) = \frac{\sum\limits_{i=0}^{m} n_i(t)}{A}$$

Then

$$F_x \equiv F_x(D), \quad x = 1, 2, \ldots, m$$

and

$$P_x \equiv P_x(D) \quad x = 0, 1, 2, \ldots, m-1.$$

Qualitatively, the relationship between F_x and D for a given age group x might take one of the following forms.

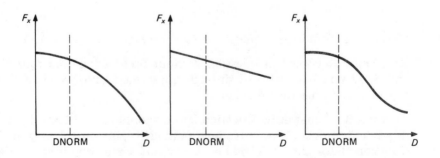

DNORM is the density which would be considered 'normal'.

Discuss these hypothetical relationships and suggest other possibilities.

Do a similar analysis for a typical P_x. Now, in each of these cases pick a simple relationship between F_x, P_x and D (for example, linear), put some values along the axes and fit the function to the resulting data.

Repeat this for each x and replace the constants in the model by the functions thus obtained. You should now have a model of the form.

$$n(t+1) = M(D)n(t).$$

Write a computer program to simulate the population structure over a period of 20 years and discuss the results obtained.

Question 3 (*Deer management policies*)

Although you may not wish to follow model 2 through in detail, the discussion on deer management policies is self-contained and you should now read it.

The basic model is easily modified to take account of various management policies. For example, a simple and frequently used policy is to remove a proportion of each age group each season. If this policy is used then the basic model may be written

$$n(t+1) = MCn(t)$$

where C is a diagonal matrix, whose diagonal entries θ_i are the proportions of deer in the ith age class which survive the cull. For example, if there is a 16% cull in age class 10 then $\theta_{10} = 0.84$. In many practical situations, the θ_i are all taken to be the same.

Now select a harvesting policy, write down the model and carry out a computer simulation to examine the effect of your policy.

Question 4 (*Influence of density effects on management policy*)

You cannot answer this question without first having done Questions 2 and 3. If you have worked through questions 2 and 3 you should find this tedious rather than difficult.

As a result of answering Question 2 you should have at least one density dependent model. Now introduce culling to this model as you did in Question 3 for the basic model. Carry out some computer simulations to examine the effects of various culling policies.

References appear on page 54.

APPENDIX 3.2

DYNAMO PROGRAM

```
*DEER POPULATION SIMULATION
NOTE
NOTE   LEVEL EQUATIONS
NOTE
1 L      CAVS.K=CAVS.J+(DT)(CBR.JK-FYRR.JK-MYRR.JK-CDR.JK)
2 L      FYRLG.K=FYRLG.J+(DT)(FYRR.JK-FMRR.JK-FYDR.JK)
3 L      MYRLG.K=MYRLG.J+(DT)(MYRR.JK-MMRR.JK-MYDR.JK)
4 L      FMATR.K=FMATR.J+(DT)(FMRR.JK-FMDR.JK)
5 L      MMATR.K=MMATR.J+(DT)(MMRR.JK-MMDR.JK)
NOTE
NOTE   RATE EQUATIONS NEXT
NOTE
6 R      CBR.KL=(BMF)(FMATR.K-FMATC.K)
7 R      CDR.KL=(CMF.K)(CAVS.K)

8 R      FYRR.KL=(FYRF.K)(CAVS.K)/2
9 R      FYDR.KL=(FYMF.K)(FYRLG.K)
10 R     MYRR.KL=(MYRF.K)(CAVS.K)/2
11 R     MYDR.KL=(MYMF.K)(MYRLG.K)
12 R     FMRR.KL=(FMRF.K)(FYRLG.K)
13 R     MMRR.KL=(MMRF.K)(MYRLG.K)
14 R     FMDR.KL=(FMMF.K)(FMATR.K)
15 R     MMDR.KL=(MMMF.K)(MMATR.K)
NOTE
NOTE   NOW THE AUXILIARIES FOLLOW
NOTE
16 A     TPOP.K=CAVS.K+FYRLG.K+MYRLG.K+FMATR.K+MMATR.K
17 A     DENS.K=TPOP.K/AREA
18 A     RAFM.K=FMATR.K/MMATR.K
19 A     RAMF.K=1/RAFM.K
20 A     REN.K=DENS.K/DNORM
20.1 A   CFAC.K=RDEN.K

21 A     FMCF.K=(NFMCF)(RAMF.K)CFAC.K)
22 A     MMCF.K=(NMMCF)(RAMF.K)(CFAC.K)
23 A     FYCF.K=(NFYCF)(CFAC.K)
24 A     MYCF.K=(NMYCF)(CFAC.K)
25 A     CCF.K=(NCCF)(CFAC.K)
26 A     FMATC.K=(FMCF.K)(FMATR.K)
27 A     MMATC.K=(MMCF.K)(MMATR.K)
28 A     FYRC.K=(FYCF.K)(FYRLG.K)
29 A     MYRC.K=(MYCF.K)(MYRLG.K)
30 A     CAVC.K=(CCF.K)(CAVS.K)
31 A     CMF.K=1-(1-NCMF)(1-CCF.K)
32 A     FYRF.K=NFYRF*(1-CCF.K)
33 A     MYRF.K=NMYRF*(1-CCF.K)
34 A     FYMF.K=1-(1-NFYMF)(1-FYCF.K)
35 A     MYMF.K=1-(1-NMYMF)(1-MYCF.K)
36 A     FMRF.K=NFMRF*(1-FYCF.K)
37 A     MMRF.K=NMMRF*(1-MYCF.K)
38 A     FMMF.K=1-(1-NFMMF)(1-FMCF.K)
39 A     MMMF.K=1-(1-NMMMF)(1-MMCF.K)
NOTE
NOTE   NOW THE INITIAL CONDITIONS ON THE LEVELS
```

```
NOTE
40  N     CAVS=96
41  N     FYRLG=33
42  N     MYRLG=32
43  N     FMATR=71
44  N     MMATR=71
NOTE
NOTE      NOW THE CONSTANTS
NOTE
45  C     AREA=2000
46  C     DNORM=.1
47  C     BMF=0.7

48  C     NCMF=.15
49  C     NFYRF=0.85
50  C     NFYMF=0.10
51  C     NMYRF=0.85
52  C     NMYMF=0.10
53  C     NFMRF=0.90
54  C     NMMRF=0.90
55  C     NFMMF=0.10
56  C     NMMMF=0.10
57  C     NFMCF=.25
58  C     NMMCF=.25
59  C     NFYCF=.25
60  C     NMYCF=.25
61  C     NCCF=.25

END

PRINT   CAVS,FYRLG,MYRLG,FMATR,MMATR,TPOP,DENS,RAFM
PRINT   CAVC,FYRC,MYRC,FMATC,MMATC
PLOT  TPOP=T(50,300)
SPEC  LENGTH=20/PRTPER=5.0/PLTPER=1.0/ITYPE=2/DT=1
```

TYPICAL DYNAMO RUNS

Normal cull is 25% per annum in all age groups.

TIME	CAVS MMATR CAVC MMATC	FYRLG TPOP FYRC	MYRLG DENS MYRC	FMATR RAFM FMATC
0.00000	96.000 71.000 36.360 26.891	33.000 303.00 12.499	32.000 0.15150 12.120	71.000 1.0000 26.891
5.0000	24.764 40.006 3.8382 6.1994	9.6047 123.99 1.4887	9.6047 0.61997E-01 1.4887	40.014 1.0002 8.2032
10.000	20.278 32.303 2.5432 4.0512	7.7240 100.33 0.96871	7.7240 0.50166E-01 0.96871	32.303 1.0002 4.0513
15.000	18.455 29.387 2.1045 3.3530	7.0290 91.278 0.80200	7.0290 0.45639E-01 0.80200	29.387 1.0000 3.3530
20.000	17.493 27.870 1.8928 3.0157	6.6660 86.564 0.72130	6.6660 0.43282E-01 0.72130	27.870 1.0000 3.0157

```
 50.00        112.5         175.0         237.5         300.0   T≡TPOP
  0- - - - - - - - - - - - - - - - - - - - - - - - - - - - - - - -
TIME+            +              +   T              +              +   TIME
    +            +          T   +                  +              +
    +            +        T      +                  +              +
    +            +     T         +                  +              +
    +            +  T            +                  +              +
    +           +T               +                  +              +
    +            T                +                  +              +
    +          T+                 +                  +              +
    +        T +                  +                  +              +
 10- - - - T - - - - - - - - - - - - - - - - - - - - - - - - - - - -
    +       T   +                 +                  +              +
    +       T   +                 +                  +              +
    +        T  +                 +                  +              +
    +      T    +                 +                  +              +
    +      T    +                 +                  +              +
    +      T    +                 +                  +              +
    +     T     +                 +                  +              +
    +     T     +                 +                  +              +
    +     T     +                 +                  +              +
 20- - -T- - - - - - - - - - - - - - - - - - - - - - - - - - - - - -
```

Normal Cull is 12% per annum in all age groups.

TIME	CAVS MMATR CAVC MMATC	FYR LG TPOP FYRC	MYR LG DENS MYRC	FMATR RAFM FMATC
0.00000	96.000 71.000 17.453 12.908	33.000 303.00 5.9994	32.000 0.15150 5.8176	71.000 1.0000 12.908
5.0000	43.051 69.518 5.5674 8.9834	16.697 215.53 2.1592	16.697 0.10777 2.1592	69.568 1.0007 9.0030
10.000	39.018 62.152 4.5195 7.1987	14.862 193.05 1.7215	14.862 0.96525E-01 1.7215	62.155 1.0000 7.1997
15.000	36.951 58.871 4.0540 6.4589	14.081 182.26 1.5449	14.081 0.91428E-01 1.5449	58.871 1.0000 6.4590
20.000	35.800 57.037 3.8054 6.0628	13.642 177.16 1.4501	13.642 0.88580E-01 1.4501	57.037 1.0000 6.0628

```
    50.00        112.5          175.0              237.5          300.0   T≡TPOP
     0- - - - - - - - - - - - - - - - - - - - - - - - - - - -
       +            +              +                +    T       +       TIME
TIME   +            +              +               +T             +
       +            +              +              T+              +
       +            +              +             T +              +
       +            +              +          T   +               +
       +            +              +        T     +               +
       +            +              +       T      +               +
       +            +              +      T       +               +
       +            +              +     T        +               +
    10- - - - - - - - - - - - - - T - - - - - - - - - - - - - - -
       +            +              +  T            +              +
       +            +              +  T            +              +
       +            +              +  T            +              +
       +            +              + T             +              +
       +            +              + T             +              +
       +            +              + T             +              +
       +            +              +T              +              +
       +            +              +T              +              +
       +            +              +T              +              +
    20- - - - - - - - - - - - - - -T- - - - - - - - - - - - - - -
```

Normal Cull is 7% per annum in all age groups.

TIME	CAVS MMATR CAVC MMATC	FYRLG TPOP FYRC	MYRLG DENS MYRC	FMATR RAFM FMATC
0.00000	96.000 71.000 10.181 7.5295	33.000 303.00 3.4996	32.000 0.15150 3.3936	71.000 1.0000 7.5295
5.0000	57.517 92.817 5.7955 9.3397	22.308 287.89 2.2478	22.308 0.14395 2.2478	92.945 1.0014 9.3782
10.000	58.900 93.814 6.0074 9.5671	22.435 291.41 2.2882	22.435 0.14571 2.2882	93.825 1.0001 9.5709
15.000	59.244 94.388 6.0791 9.6851	22.577 293.17 2.3166	22.577 0.14659 2.3166	94.389 1.0000 9.6855
20.000	59.465 94.742 6.1247 9.7580	22.661 294.27 2.3340	22.661 0.14714 2.3340	94.742 1.0000 9.7580

```
50.00        112.5         175.0          237.5          300.0   T≡TPOP
   0- - - - - - - - - - - - - - - - - - - - - - - - - - -
     +            +             +              +          T  +      TIME
TIME +            +             +              +          T  +
     +            +             +              +           T  +
     +            +             +              +           T  +
     +            +             +              +           T  +
     +            +             +              +           T  +
     +            +             +              +            T +
     +            +             +              +            T +
     +            +             +              +            T +
  10- - - - - - - - - - - - - - - - - - - - - - - - - - T -
     +            +             +              +            T +
     +            +             +              +            T +
     +            +             +              +            T +
     +            +             +              +            T +
     +            +             +              +            T +
     +            +             +              +            T +
     +            +             +              +            T +
     +            +             +              +            T+
     +            +             +              +            T+
  20- - - - - - - - - - - - - - - - - - - - - - - - - -T-
```

REFERENCES

General Ecological Texts

1. E.J. Kormondy   *Concepts of Ecology* (New Jersey, Prentice-Hall, 1969)
2. K.E.F. Watt   *Ecology and Resource Management* (New York, McGraw-Hill, 1968)

Deer Management

3. J.R. Beddington   Economic and Ecological Analysis of Red Deer Harvesting in Scotland, *J. Environmental Mgmt.* **3**, 91–103 (1975)
4. V.P.W. Lowe   'Population Dynamics of the Red Deer *(Cervus Elaphus L)* on Rhum', *J. Anim. Ecol.* **38**, 425–57 (1976)
5. *The Red Deer Commission Annual Report for 1976* Edinburgh, HMSO, 1977

Techniques

6. R.G. Coyle   *Management System Dynamics* (London, J. Wiley, 1977)
7. J.W. Forrester   *Industrial Dynamics* (Cambridge, Mass., MIT Press 1961)
8. J.W. Forrester   *Urban Dynamics* (Cambridge, Mass., MIT Press, 1969)
9. J.W. Forrester   *World Dynamics* (Cambridge, Mass., Wright Allen Press, 1971)
10. M.R. Goodman   *Study Notes in System Dynamics* (Cambridge, Mass, Wright Allen Press, 1974)
11. J.H. Pollard   *Mathematical Models for the Growth of Human Populations* (London, Cambridge University Press, 1975)
12. A.L. Pugh   *DYNAMO II User's Manual* (Cambridge, Mass., MIT Press, 1973)

Methodology

13. J. Lighthill Presidential address, *Developments in Mathematical Education* ed. A.G. Howson (London, Cambridge University Press, 1973) p. 88
14. D.J. White   *Decision Methodology* (London, Wiley, 1975)

4.

Modelling the Leaching of Nitrates in Fallow Soils

A.O. Moscardini

4.1 STATEMENT OF THE PROBLEM

One of Britain's most important crops is wheat. In a typical year about three million acres will be sown and the harvest would be worth about five hundred million pounds. One of the many farmland costs, along with costs of seed, land, plant machinery and labour, is that of the fertilizers. Various types of fertilizer are used but the cost of nitrate-based fertilizers alone is of the order of sixty million pounds. Any saving here would obviously be desirable and this is the object of this case study.

As nitrates are soluble, they are affected by the movement of water in the soil. This process is called 'leaching'. In dry spells the nitrates tend to remain around the surface and with heavy rainfalls they could be leached out of the soil completely — see Table 4.1. As most roots tend to be within two or three feet of the surface, any nitrates leached below this distance can be considered lost. Usually a first dressing is applied in early autumn when the seed is planted and a second, larger dressing is applied in the spring. It is important that the farmer estimates the amount of this second dressing accurately. Two little at this stage will have obvious effects; too much nitrate also affects the yield because of excess salinity in the soil. If, for example, nine farmers out of ten estimate correctly and only one gets it wrong the wastage involved could amount to about six million pounds.

Thus, the movement of water in soil and hence the distribution of surface applied nitrates is an important area of study. Can a mathematical model be constructed that would help the soil scientists to investigate this problem?

Some typical figures for Britain are as follows:

Crop	Wheat.
Acreage	2.5–3 million cultivated per year depending mainly on winter weather conditions.
Yield	1.5–2.5 tonnes/acre.
Value	Approximately £100 per tonne.
Cost	Approximately £3 per 15 kg.
Rainfall	The following figures are for Oxfordshire, but there would usually be greater variations between Winter and Summer throughout the country.

	Winter (mm)	Summer (mm)
1966	271	404
1967	367	382
1968	319	520
1969	348	308

Application of Nitrates   20–100 kg per acre.

Fig. 4.1   Response curves for wheat crop

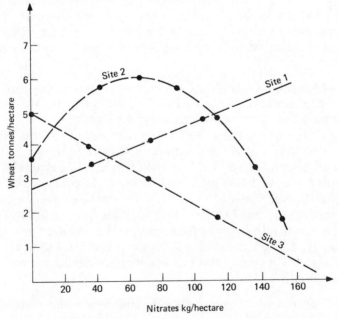

Fig. 4.1 shows some response curves for wheat crops. Three sites were chosen and all behaved differently. Site 1 had heavy leaching and so more nitrates were continually needed to encourage yield. Site 2 reached a desired level of nitrates at about 80 kg/hectare and thence excess salinity reduced the yield. Site 3 suffered little leaching and nitrates built up in the soil reducing the yield. Varying amounts of rainfall will produce more or less leaching and thus can be thought of as shifting figures horizontally to the left or right.

4.2 BACKGROUND TO THE PROBLEM

Soil, as is commonly known, could be described as the weathered and fragmented outer layer of the Earth's land surface, which has been formed from the disintegration and decomposition of rocks by physical and chemical process and which has been influenced by the activity and accumulated residues of numerous biological species. The three phases of nature exist in it: the solid phase consisting of soil particles, the liquid phase consisting of soil water, and the gaseous phase consisting of soil air.

The solid phase consists of particles which differ in chemical and mineralogical composition as well as in size, shape and orientation. The mutual arrangement of these particles in the soil determines the characteristics of the pore spaces in which water and air are transmitted or retained. These pore spaces can amount to up to 50% of the total volume.

The chemical behaviour of the soil is determined mainly by the colloidal clay. Clay particles absorb moisture and swell/shrink on wetting/drying, obviously affecting the total pore space. Most of them are also negatively charged and therefore there is physiochemical activity as well.

The structure of soil is also important. In general, it is possible to recognise three types of soil structure:

(a) single grained (when the particles are completely unattached to each other);

(b) massive (when the particles are bonded in large and massive blocks); and

(c) an intermediate condition in which the soil particles are arranged in small units known as 'aggregates' or 'pods'.

Thus it can be seen that soil is an exceedingly complex system.

4.2.1 THE STATE OF WATER IN THE SOIL

Soil water, like other bodies in nature, can contain energy in different forms, namely kinetic and potential. Since the movement of water in the soil is quite slow, kinetic energy (proportional to the velocity squared) is quite small and therefore is considered negligible, but potential energy (due to position or internal condition) is of primary importance. Soil water moves in the direction of decreasing potential energy. It is not therefore the amount of potential energy 'contained' in the water that is important but the relative level of that energy in different regions within the soil. The concept of 'soil water potential' θ, is a yardstick for this energy. It expresses the potential energy of soil water relative to that of water in a standard reference state. This can be measured *in situ* using an instrument called a 'tensiometer'.

When the soil is saturated and its water is at a hydrostatic pressure greater than the atmospheric pressure, $\theta > 0$ and water will flow out into a drainage reservoir. If the soil is moist but unsaturated, its water will no longer be free but is generally constrained by capillary and absorptive forces; θ is then usually negative, and the soil draws any water in contact with it as a blotter draws ink. Under normal conditions in the field, the soil is generally unsaturated and the soil water potential, called the matric or suction potential, is negative. The latter is measured using pF (that is, the logarithm of the negative pressure head in centimetres of water).

The relationship between soil water content and soil potential varies considerably for different soils — see Fig. 4.2(a). A function known as the 'soil moisture characteristic' shows this graphically but no equation as yet exists. The absorption and pore effects are often too complex and it also depends strongly on soil texture and soil structure — see Figs. 4.2(b) and (c). It is also complicated by the effect called hysteresis. If the relationship between θ and water content is plotted as the soil is drying (desorption) and then as the curve is wetting (sorption), the two curves, it can be seen, are not identical. The equilibrium wetness at a given θ is greater in desorption than in sorption. This dependence of the equilibrium content and state of soil water upon the direction of the process leading up to it is called hysteresis.

Soil has a minimum and maximum storage capacity. The 'field capacity' is a measure of the maximum water storage. This is the presumed water content at which internal drainage ceases. The minimum amount of water retained by the soil is that retained by surface tension alone. This is important as upon it depends, to some extent, the power of the soil to resist drought by retaining water for the crop between intervals of rain. This is called the 'evaporation limit'.

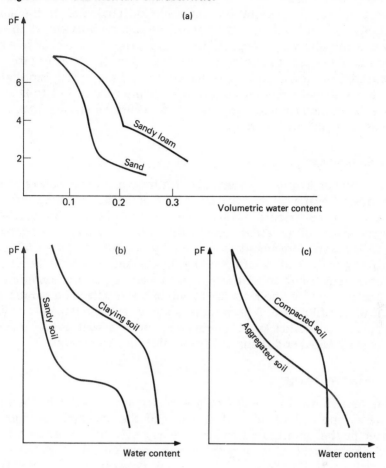

Figure 4.2 Soil moisture characteristics

(a)

pF

6

4

2

0.1   0.2   0.3   Volumetric water content

Sandy loam

Sand

(b)

pF

Sandy soil   Claying soil

Water content

(c)

pF

Compacted soil   Aggregated soil

Water content

4.2.2 NORMAL CYCLE OF WATER IN THE FIELD

This normally follows the pattern:

Infiltration — Internal drainage — Groundwater drainage — Evaporation — Uptake of soil water by plants

A brief description of each stage is given below.

Infiltration

This is the process by which the water enters the soil. This is mainly by rainfall and the equivalent problem is when the water lies on the soil and run-off occurs (important for erosion). For downward flow there are two forces acting, the gravitational and suction gradients. If the surface is initially dry and suddenly saturated, the suction gradient at the surface layer is initially very steep. As the water penetrates deeper this gradient decreases and the filtration rate settles down to a steady rate, gravity-induced. The surface run-off is that portion of rain which is not absorbed by the soil and does not accumulate on the surface but runs away and collects in gullies and streams. This occurs when the rain intensity exceeds the infiltration rate.

Internal drainage

When rain or irrigation ceases, the infiltration process comes to an end. Downward water movement does not cease immediately but internal drainage or redistribution takes place. The typical moisture profile at this stage consists of a wetted zone in the upper regions and a relatively dry zone beneath. Movement is caused by suction or potential gradients and/or gravity and is affected by hysteresis, which in general works against redistribution. Different soils vary greatly — sometimes by a factor of ten — in their ability to retain water after comparable periods following infiltration. Soil wetness at any given time depends on the soil's hydraulic properties and layering sequence as well as the quantity of water infiltrated and the initial (pre-infiltration) wetness.

Groundwater drainage

This term normally applies to the movement of water in the saturated zone as opposed to internal drainage of the unsaturated zones. Whilst water in the unsaturated soil is strongly affected by suction gradients and its movement is subject to very considerable variations in conductivity resulting from changes in soil wetness, groundwater is always under positive hydrostatic pressure and hence it saturates the soil. Thus no suction gradients and no variations in wetness or conductivity occur below the water table (defined as the locus of points at which the hydrostatic pressure equals the atmospheric pressure) and the hydraulic conductivity is fairly constant in time and space.

Evaporation

This can take place from plants (called transpiration — dealt with later) and also from the soil surface (in the absence of vegetation). If the top layer of the soil is initially wet the process of evaporation will generally reduce soil wetness and then increase suction at the surface. This in turn will cause soil water to be drawn upward from the layers below. This rise of water in the soil from a free water surface (i.e. water table) has been termed capillary rise from the obvious analogy. The rate of evaporation from a drying soil is determined by external meteorological conditions and by the moisture-supplying/transmitting properties of the soil profile. When the soil is relatively wet and of relatively high conductivity, its ability to deliver water to the evaporation zone is not likely to be limiting and the actual evaporation rate will tend to equal the potential rate determined by the meteorological conditions. As the soil dries, its ability to transmit and supply water is reduced and the evaporation rate falls. Water will never be completely evaporated from a section of soil as there will always remain a minimum amount of water. This amount is dependent on the type of soil considered and, as stated previously, is known as the evaporation limit.

Uptake of soil water by plants

The state of movement of water in the soil, plant and atmosphere is affected by a complex set of interactions and processes which occur simultaneously at different rates. The first link in the chain is the flow of water in the unsaturated soil surrounding the root and as such it can affect the overall flow process. The ability of a plant to obtain sufficient water supply depends not only on the amount and potential of soil water but also on the required flow rate and transmitting properties of the soil.

Plant response to the soil moisture regime depends also on a large number of environmental variables not included in this discussion. Soil factors such as aeration, nutrient availability and mechanical properties, plant factors such as age and various genetic traits and meteorological factors such as light intensity, day length and atmospheric composition can influence not only the magnitude of plant response to soil water but also the direction of this response.

4.3 EXPERIMENTAL DATA

Tables 4.1 and 4.2 summarise the results of a controlled experiment by an agricultural research establishment. A careful monitoring of a test field was conducted as follows: the field soil was examined at irregular intervals and the amount of nitrates present at successive vertical levels was chemically determined. Twelve successive levels, each 300 mm thick were examined each time and the daily rainfall and evaporation were also noted. It can be seen that after 18 days there had been heavy evaporation and little rainfall so the nitrates had been leached upwards, but by the 130th day no nitrates were left in the soil. Typical values for the field capacity and the evaporation limit measured (as a percentage by volume) during this experiment are 26% and 14%.

Table 4.1

Strata	Nitrates (parts per million)						
	Initial	18 days	32 days	60 days	81 days	130 days	159 days
1 (Top)	127.858 ± 17	161.316 ± 30	111.32 ± 15	139.38 ± 14.29	15.531 ± 2.69	0.00 ± 0.00	0.542 ± 0.542
2	86.692 ± 15	33.907 ± 2.7	83.00 ± 10.8	55.34 ± 10.4	26.297 ± 5.56	0.00 ± 0.00	0.00 ± 0.00
3	71.830 ± 14	32.061 ± 3.4	30.403 ± 4.9	28.819 ± 4.4	35.311 ± 5.773	0.00 ± 0.00	0.00 ± 0.00
4	30.442 ± 10	11.393 ± 5.53	11.959 ± 4.9	12.30 ± 4.5	40.263 ± 4.969	0.00 ± 0.00	0.00 ± 0.00
5	0.000 ± 0.00	0.000 ± 0.00	0.439 ± 0.439	4.47 ± 2.46	29.85 ± 8.815	0.00 ± 0.00	0.00 ± 0.00
6	0.000 ± 0.00	0.000 ± 0.00	0.00 ± 0.00	4.25 ± 0.95	33.988 ± 10.46	0.00 ± 0.00	0.00 ± 0.00
7	0.753 ± 0.753	0.016 ± 0.016	0.00 ± 0.00	0.242 ± 0.242	14.157 ± 4.103	0.00 ± 0.00	0.00 ± 0.00
8	2.160 ± 2.16	0.041 ± 0.041	0.00 ± 0.00	0.00 ± 0.00	7.687 ± 3.79	0.00 ± 0.00	0.00 ± 0.00
9	0.372 ± 0.372	3.047 ± 2.25	0.00 ± 0.00	0.939 ± 0.939	16.786 ± 4.776	0.00 ± 0.00	0.00 ± 0.00
10	0.587 ± 0.587	0.964 ± 0.43	0.00 ± 0.00	5.096 ± 2.564	23.394 ± 2.135	0.00 ± 0.00	0.00 ± 0.00
11	0.00 ± 0.00	6.353 ± 3.57	0.00 ± 0.00	2.199 ± 1.72	13.889 ± 4.679	0.00 ± 0.00	0.00 ± 0.00
12 (Bottom)	0.00 ± 0.00		0.00 ± 0.00	2.594 ± 2.594	15.657 ± 3.72	0.00 ± 0.00	0.00 ± 0.00
Total Nitrate in profile	320.694 ± 28.560	249.098 ± 31.18	237.121 ± 19.74	255.629 ± 19.38	272.810 ± 19.441	0.00 ± 0.00	0.542 ± 0.542
Total rainfall in period		0.127	1.29	2.66	4.801	10.337	0.717
Total evaporation in period		1.996	1.738	3.761	3.77	6.126	0.873

Table 4.2 Daily evaporation (cm) and rainfall (cm) for given period

Day	Evap.	Rain	Day	Evap.	Rain	Day	Evap.	Rain	Day	Evap.	Rain	Day	Evap.	Rain
1	0.206	0.051	33	0.037	0.00	65	0.031	0.00	97	0.16	0.01	129	0.091	0.00
2	0.223	0.000	34	0.029	0.00	66	0.236	0.864	98	0.00	4.216	130	0.091	0.00
3	0.128	0.00	35	0.297	0.711	67	0.236	0.00	99	0.00	0.152	131	0.183	0.00
4	0.128	0.00	36	0.366	0.00	68	0.229	0.025	100	0.069	0.406	132	0.015	0.00
5	0.288	0.00	37	0.404	0.00	69	0.457	0.00	101	0.069	0.305	133	0.024	0.00
6	0.055	0.00	38	0.054	0.00	70	0.043	0.127	102	0.069	0.00	134	0.015	0.00
7	0.165	0.051	39	0.054	0.00	71	0.119	0.229	103	0.091	0.00	135	0.02	0.00
8	0.206	0.00	40	0.037	0.057	72	0.122	0.762	104	0.160	0.00	136	0.02	0.00
9	0.165	0.00	41	0.320	0.279	73	0.122	0.076	105	0.137	0.00	137	0.02	0.00
10	0.110	0.00	42	0.049	0.00	74	0.122	0.787	106	0.206	0.00	138	0.006	0.00
11	0.045	0.00	43	0.052	0.00	75	0.091	0.940	107	0.146	0.00	139	0.04	0.127
12	0.018	0.00	44	0.036	0.00	76	0.069	0.00	108	0.146	0.00	140	0.024	0.00
13	0.070	0.00	45	0.0267	0.66	77	0.320	0.00	109	0.146	0.00	141	0.018	0.00
14	0.018	0.00	46	0.267	0.102	78	0.297	0.00	110	0.274	0.00	142	0.025	0.00
15	0.030	0.025	47	0.434	0.00	79	0.030	0.00	111	0.052	0.00	143	0.025	0.076
16	0.040	0.00	48	0.32	0.33	80	0.030	0.00	112	0.023	0.33	144	0.075	0.000
17	0.040	0.00	49	0.183	0.051	81	0.034	0.00	113	0.137	0.00	145	0.018	0.051
18	0.040	0.00	50	0.114	0.102	82	0.034	0.00	114	0.229	0.00	146	0.024	0.00
19	0.024	0.00	51	0.025	0.00	83	0.043	0.00	115	0.03	0.00	147	0.024	0.00
20	0.058	0.00	52	0.025	0.00	84	0.034	0.102	116	0.03	0.00	148	0.012	0.00
21	0.024	0.00	53	0.025	0.00	85	0.069	0.029	117	0.114	0.279	149	0.001	0.00
22	0.052	0.00	54	0.027	0.00	86	0.244	0.00	118	0.343	0.991	150	0.001	0.00
23	0.051	0.00	55	0.058	0.102	87	0.032	0.00	119	0.16	0.254	151	0.001	0.00
24	0.051	0.00	56	0.052	0.178	88	0.244	0.229	120	0.206	0.152	152	0.006	0.00
25	0.051	0.00	57	0.024	0.00	89	0.040	0.00	121	0.091	0.254	153	0.023	0.127
26	0.058	0.00	58	0.064	0.076	90	0.030	0.00	122	0.091	0.356	154	0.023	0.00
27	0.030	0.00	59	0.064	0.00	91	0.046	0.00	123	0.091	0.660	155	0.069	0.00
28	0.046	0.00	60	0.064	0.00	92	0.137	0.279	124	0.183	0.00	156	0.015	0.00
29	0.480	1.29	61	0.32	0.889	93	0.168	0.00	125	0.137	0.152	157	0.002	0.00
30	0.381	0.00	62	0.343	0.102	94	0.168	0.381	126	0.091	0.00	158	0.015	0.178
31	0.381	0.00	63	0.343	0.00	95	0.168	0.00	127	0.274	0.00	159	0.183	0.178
32	0.051	0.00	64	0.183	0.00	96	0.411	0.00	128	0.091	0.00			

4.4 ANTICIPATED RESPONSE TO THE BACKGROUND INFORMATION

When the background provided has been read, the full complexity of the problem will be apparent. There is obviously much more information supplied than is necessary and the immediate task is to decide the relevant and essential sections. This is an important part of the modeller's art and decisions taken at this stage determine the type of model that will arise. It is wise to keep the first model as simple as possible. Then modifications can always be made to deal with any desired complications.

Ths most important fact contained in the background is that the movement of nitrates is directly associated with the movement of water in the soil and this can be both in an upward or downward direction.

If the soil water potential θ is now used, the approach will lead eventually to some form of the diffusion equation which could be solved numerically. The suitability of this approach will depend upon the level of the students whilst from the practical point of view it also needs values of certain constants which are not always readily available. For these reasons, this approach is not recommended as the initial one but could be developed as an extension to the problem. It has been covered in detail by various authors especially Gardner and Sheidigger.[1-3]

Two other concepts in the background are those of field capacity and evaporation limits, which can be identified as the maximum and minimum levels of water contained in a certain section of soil. Discussion around these concepts will naturally lead to the question of how to deal with a section of soil which is not necessarily, indeed hardly ever, a homogenous entity. It would be hoped that the students would suggest treating the soil, therefore, as a succession of layers each with its own individual composition which would be typified by the field capacity and evaporation limit. This approach which was first taken by Burns[4,5] and has also been studied by Terkeltoub and Addiscott[7-9] leads to a much simpler model and is the one used by the author.

Further complications that were mentioned in the background are now viewed as extensions to the problem. They can be treated as they arise and added to the original model.

These are not the only two approaches that can be adopted but they are the two that seem to arise from the background notes. Of course, there are many other approaches and relationships which a group may suggest. Some will lead to dead-ends and some to valid models.

4.5 AN AGREED APPROACH TO THE PROBLEM

The soil is to be treated as a succession of separate layers, each characterised by its own field capacity, evaporation limit, water content and soil content. External variables are the daily rainfall and evaporation which can both be measured experimentally without much difficulty. Considerations such as water potential, hysteresis, pore spacing, cracks and gullies are ignored. A simple algebraic model is constructed that treats each layer analogously to a tank of water which has prescribed maximum and minimum limits. The model is to account for the movement (both upwards and downwards) of water through the layers and thus calculate the amount of nitrates leached through the soil.

4.6 SIMPLE MODEL OF LEACHING

This model uses two concepts that must be fully understood, namely:

(a) *Field Capacity*. This is a long used term that describes the presumed water content at which internal drainage ceases, i.e. it is the maximum water storage capacity of a soil under field conditions.

(b) *Evaporation Limit*. When the soil is drying, water is being lost to the atmosphere. It is assumed that this will stop at a minimum water content called the evaporation limit.

The evaporation limit and field capacity can then be regarded as the minimum and maximum levels of water content in the soil. This simple model is discrete and treats the soil as a succession of separate layers, each characterised by its own field capacity F, evaporation limit L, initial water content W_0 and initial salt content S_0. It uses the simple equation that the net water X supplied to or removed from the soil each day is connected with the rainfall and the evaporation by

$$X = R \text{ (Rainfall)} - E \text{ (Evaporation)}.$$

If $X > 0$ rainfall exceeds evaporation so that downward leaching occurs.

If $X = 0$ no movement of water or salts occur.

If $X < 0$ evaporation is dominant and upward movement of water and salts takes place.

The model developed will assume that

(1) Net water supply is applied to the top layer causing a temporary increase in water content. Infiltration rates are sufficiently high to avoid run-off.

(2) The added water remains in the segment for a short time so that mixing occurs, i.e. any salt in the rainwater T mixes completely with that already in the layer S during this period and the new salt content becomes $(S + T)$.

(3) If the new water content of the layer is less than F, no further redistribution occurs.

(4) If the water content now exceeds F, the field capacity, the excess water is simply transferred to the layer beneath.

We can use a simple water balance equation

$$\text{Water lost } (W_p) = X + W_0 - F$$

to evaluate the water lost from each segment. We can then calculate the fraction Z_p of the water that was temporarily held in the segment,

$$Z_p = \frac{W_p}{X + W_0}$$

Since mixing was assumed to have occurred after the water was applied, Z_p also represents the fraction of salt that was transferred downwards by leaching. Therefore the actual quantity leached from the segment was $(S + T)Z_p$, whilst the salt remaining in the segment is $(S + T)(1 - Z_p)$. The values of W_p and the amount of salt leached become X and T for the next segment and the whole procedure is repeated until infiltration ceases.

If $X < 0$ evaporation exceeds the rainfall applied and a different routine is used.

The water available for evaporation W_E is calculated for each segment in turn where $W_E = W_0 - L$ and the topmost segment having $W_E > 0$ is located. Water will then be removed from this section first. If the net evaporative requirement is small, then this segment can probably supply all the water required. If not the difference has to be made up with water derived from lower segments.

The fraction of water lost by capillary rise from a segment is

$$Z_c = \frac{W_c}{W_0}$$

where W_c is the amount of water arising from the segment, and thus actual salt lost is $S_0 Z_c$. After the transfer the final water content of the segment W_f becomes

$$W_f = W_0 - W_c$$

The water and salts lost from this segment move into the one immediately above where they temporarily increase the water and salt contents of this region. While water and salt are present in this new segment, mixing is again assumed to occur, so the rising salt and water are added to the original salt and water contents before the losses from this new region are calculated. The routine is applied progressively until the water is lost from the surface. The calculations are then repeated for the next top-most layer with $W_E > 0$ until the total required daily evaporation is completely satisfied. As soon as the actual water balances the required loss, the calculations are stopped and no further redistributions are assumed to occur.

4.7 TESTING THE MODEL

The model was programmed in BASIC for use on a DEC PDP 11/40. Fig. 4.3 shows a flow chart for the program which is listed in Appendix 4.1.

The data input is contained in two files PUT1.DAT and PUT2.DAT. The first contains values of the rainfall and evaporation over a number of days and the second contains an initial profile of the soil. This consists of values of the field capacity, evaporation limit, mean moisture content and mean nitrate content for each layer. The output consists of values of the mean moisture and mean nitrate content every fifth day followed by a rough bar chart of the percentage nitrate distribution which is plotted on the terminal.

Various runs of the model were made and upward and downward move-ment of the nitrates was observed. The results for one run are shown in Appendix 4.2; the input for this run being taken from the experimental data in Tables 4.1 and 4.2. Day 1 of the run corresponds to day 60 in Table 4.1 and the values of rainfall and evaporation for the next 40 days were used. The initial profile was again that of day 60, and this is shown in Table 4.3. (Variations of field capacity and evaporation limit were allowed from layer to layer. The values were thought to model the physical behaviour of the soil and reflect the intuition of the modeller!) The results of Appendix 4.2 show that the nitrates were almost completely leached out of the soil by day 40 and in fact vanished by day 49. This corresponds to the experimental data in Table 4.1 where they had disappeared by day 130. Another comparison can be made between day 20 of the run and day 81 from Table 4.1. It can be seen that there is a similar overall behaviour in the movement of the nitrates. This is very encouraging as the program has scope for considerable improvement especially with regard to evaporation calculations. It should not, however, be taken as a final model but rather as a basis for the development of a good computer model.

Fig. 4.3 Flowchart for leaching program

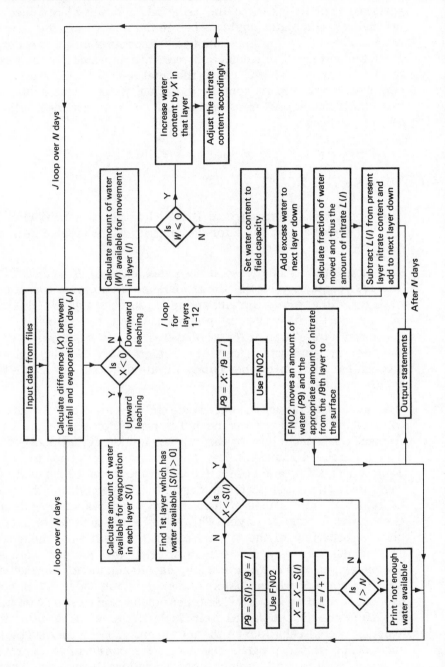

Table 4.3 Contents of file PUT2.DAT (Profile for day 60)

		Field capacity	Evaporation limit	Mean moisture content	Mean nitrate content
Level	1	0.24	0.02	139	56.96
	2	0.24	0.05	55	22.54
	3	0.25	0.14	29	11.88
	4	0.26	0.14	12	4.98
	5	0.29	0.15	5	2.04
	6	0.29	0.15	4	1.64
	7	0.29	0.15	0	0
	8	0.29	0.16	0	0
	9	0.29	0.16	0	0
	10	0.29	0.14	0	0
	11	0.29	0.14	0	0
	12	0.29	0.13	0	0

4.8 APPLICATION OF THE MODEL

The model was tested against the experimental data as discussed before and a similar leaching pattern was observed. A more theoretical application of the model is as follows.

Assume that the fraction of nitrate leached from a segment, calculated as

$$Z = \frac{W_p}{W_0 + X}$$

is constant for all segments.

Thus if the amount of nitrate in layer 1 before leaching is S_1, then

amount leached from layer 1 is $S_1 Z$,

amount in layer 2 before leaching is $S_2 + S_1 Z$,

amount leached from layer 2 is $S_2 Z + S_1 Z^2$,

and amount leached from layer r is $S_r Z + S_{r-1} Z^2 + \ldots + S_1 Z^r$.

If the nitrate is applied to the top segment only, then $S_2 = S_3 = S_4 \ldots = S_r = 0$, $S_1 = S$, and the total amount of nitrate leached from layer r is SZ^r. The fraction f that has been leached below layer r is

$$f = \frac{SZ^r}{S} = Z^r = \left(\frac{W_p}{W_0 + X} \right)^r.$$

However, $W_0 + X = W_p + F$, where $F = Vt/100$, V being the percentage volumetric water content and t the segment thickness. Thus,

$$f = \left(\frac{W_p}{W_p + Vt/100} \right)^r$$

which, on letting $h = rt$, gives

$$f = \left(\frac{W_p}{W_p + Vt/100}\right)^{h/t}.$$

This is based on the assumption that the solution of a given layer is in equilibrium with its drainage water at all times, which is most likely to be true if t is small. Provided t is small, then small changes in its value will only have small effects on f. Thus, assuming $t = 1\,\mathrm{cm}$ we have that

$$f = \left(\frac{W_p}{W_p + V/100}\right)^{h}.$$

This can be used to calculate the fraction of surface applied nitrate leached below any depth h cm in soil of field capacity V (% by volume) after W_p cm of water has drained through the system. W_p could be estimated directly from the difference between the rainfall and the evaporation when the soil is initially at field capacity.

By taking logs and expanding using the Maclaurin series for $\ln(1 + x)$ we have

$$\ln f = \frac{-Vh}{100\,W_p} \quad \text{if} \quad \frac{V}{100} \ll W_p.$$

If $f = 0.5$, $h = h_{\frac{1}{2}}$ (depth which contains half the nitrate), then

$$h_{\frac{1}{2}} = \frac{100\,W_p \ln 0.5}{V} \doteq \frac{-70\,W_p}{V}$$

This compares well with a well established formula by Rousselle–Levin

$$d = \frac{-100\,W_p}{V}$$

where d is the depth of peak nitrate concentration which approximated to the mean displacement in many of their experiments.

4.9   DISCUSSION OF THE MODEL

It was found that the choice of segment size and frequency with which the algorithm was used influenced the amount of leaching and evaporation predicted by the model. The model assumes that total mixing occurs in each segment and this may be invalid if very large segments and large values of rainfall and evaporation are used. In theory, one should thus keep the segment size small and use daily readings.

All diffusive movement is ignored in the model, but it is hoped that the errors occurring from this are more than compensated by the gain in simplicity.

Cracks and aggregates in the soil are also disregarded by the model. These can lead to preferred pathways for the water through the soil and thus nitrates could be leached away far quicker than the model would allow.

However the model, although it has limitations, can be applied to freely drained soils (the sandy loam type especially) and gives reasonable approximation of the position of nitrates without involving any complicated or obscure soil parameters.

4.10   EXTENSIONS TO THE MODEL

1.   An obvious extension would be to allow for soils with physical irregularities such as cracks or aggregates. This could be done by creating some sort of 'fissure factor'. An investigation, generally into the pore spacing gives good scope for modelling. Suggested problems are the following:

1a.   The mutual arrangements of particles in the soil determine the characteristics of the pore spaces in which water and air are transmitted or retained. These pore spaces can amount up to 50% of total volume. Investigate.

1b.   The 'specific surface' of a soil can be defined as the total surface area (both internal and external) of particles per unit volume of dry soil. Investigate.

2.   Hysteresis could be taken into account, i.e. the different water holding capacity of the soil depending on whether it is wetting or drying. This could be modelled using the so called 'ink-bottle' effect.

3.   Models using more advanced mathematics can be developed to account for the diffusion effect. Most of these will involve physical laws such as D'Arcy's law and will produce some version of the classical diffusion equation. This could then be solved numerically and the results compared with the simpler model.

4.   Roots are a main source of water removal from the soil. They could be represented by concentric cylinders and a model could be created that simulated the removal of soil water through the roots. This could then be a useful extension to the previous models.

5.   Explain the action of silica gel as a desiccant. (*Harder*)

References appear on page 76.

APPENDIX 4.1   COMPUTER LISTING

COMPUTER LISTING

```
5 DIM F(20),L(20),A(20) ,M(20),R(200),E(200),S(20)
6 DIM L1(20)
7 DIM P(12),P1(12)
9!    *****DATA INPUT BY FILES*****
10 OPEN"PUT1.DAT" AS FILE 1%
12 INPUT#1% N
14 INPUT#1% F(I):L(I):A(I):M(I) FOR I=1 TO N
16 OPEN "PUT2.DAT" AS FILE 2%
18 INPUT#2%.N1
20 INPUT#2% R(I):E(I) FOR I=1 TO N1
22 CLOSE 1%.2%
23!*** F(I)IS FIELD CAPACITY,L(I) IS EVAPORATION LIMIT
24!*** M(I)=IS WATER CONTENT,A(I)  IS NITRATE CONTENT
25!***R(I) IS RAINFALL ON ITH DAY,E(I) IS EVAPORATION ON ITH DAY
26!***** END OF DATA INPUT*****
27 GOTO900
50 FOR J=1 TO N1
60 T=0
70 X=R(J)-E(J)
80 IF X<=0 THEN 250
85!*********DOWNWARD LEACHING**********
90 FOR I=1 TO N
100 W=X+M(I)-F(I)
110 IF W<=0 THEN 190
120 M(I)=F(I)
122 IF I=N THEN 130
130 Z=W/(W+F(I))
140 A(I)=A(I)+L1(I-1)
150 L1(I)=A(I)*Z
160 A(I)=A(I)-L1(I)
165 IF A(I)<0 THEN A(I)=0
170 X=W:T=L1(I)
180 NEXT I
185 GOTO 350
190 M(I)=M(I)+X
200 A(I)=A(I)+L1(I-1)
210 GOTO  350
245!********** UPWARD LEACHING**********
250 S(I)=M(I)-L(I) FOR I-1 TO N
251 FOR I=1 TO N
252 IF S(I)<0 THEN S(I)=0
253 NEXT I
260 I=1:T=0
261 X=-X
270 IF S(I)<=0 THEN I=I+1:IF I>(N-1) THEN 343 ELSE 270
280 IF X S(I) THEN 330
285 X=X-S(I)
290 P9=X:I9=I
295 P=FN02
320 I=I+1
322 IF I>N THEN  343
325 GOTO 270
330 P9=X:I9=I
340 P=FN02
343 !
```

```
345 &:&
346 B=0
350 J1=J/5-INT(J/5)
351 IF J1=0 THEN 900
355 NEXT J
360 GOTO 900
650 !***THIS FN TAKES AMOUNT P9 OF WATER AND PROP.AMOUNT OF SALT***
651 !***FROM ITH LEVEL TO THE SURFACE                          ***
660 DEF FNO2
665 IF I9=1 THEN 780
670 T=0
680 Z=P9/M(I9)
690 A(I9)=A(I9)+T
700 T=A(I9)*Z
710 A(I9)=A(I9)-T
720 M(I9)=M(I9)-P9
725 IF M(I9)<L(I9) THEN M(I9):P9=P9-S(I9)
730 I9=I9-1
740 M(I9)=M(I9)+P9
745 IF M(I9)>F(I9) THEN M(I9)=F(I9)
750 IF I9=1 THEN 780
760 GOTO 680
780 M(I9)=M(I9)-P9
785 IF M(I9)<L(I9) THEN M(I9)=L(I9):P9=P9-S(I9)
790 A(I9)=A(I9)+T
800 FNEND
900 &:&:&:&:&:&:&:&"TIME AFTER TREATMENT =":J: "DAYS"
910 &
920 PRINT "STRATA": TAB(16):"MEAN MOISTURE":TAB(36):"MEAN NITRATE":TAB(62):"% NO.
925 PRINT TAB(2):"NO.":TAB(19):"CONTENT":TAB(39):"CONTENT":TAB(59):"CONCENTRATE"
927 B=B+A(I) FOR I=1 TO 12
928 P(I)=A(I)*100/B FOR I=1 TO 12
930 FOR I=1 TO N
940 PRINT TAB(2):I:TAB(20):M(I):TAB(39):A(I):TAB(59):P(I)
950 NEXT I
951 P1(I)=INT(P(I)/2.5) FOR I=1 TO 12
952 &:&:&:&:&:&TAB(28):"% OF NITRATE PRESENT "
953 PRINT TAB(28):"   10  20  30  40  50  60"
954 FOR I=1 TO 12
955 PRINT TAB(19): "LEVEL":I
956 PRINT TAB(28):"*" FOR J9=1 TO P1(I)-1
957 IF P1(I)=0 OR P1(I)=1 THEN &" " ELSE &"*"
958 NEXT I
959 &:&:&:&:&:&:&
960 IF J=N1 THEN 1000 ELSE 355
1000 END
```

APPENDIX 4.2   COMPUTER RESULTS

APPENDIX 4.2   COMPUTER RESULTS

TIME AFTER TREATMENT = 0 DAYS

STRATA NO.	MEAN MOISTURE CONTENT	MEAN NITRATE CONTENT	% NITRATE CONCENTRATION
1	.06	139	56.9672
2	.12	55	22.541
3	.14	29	11.885
4	.14	12	4.91803
5	.15	5	2.04918
6	.15	4	1.63934
7	.15	0	0
8	.16	0	0
9	.16	0	0
10	.14	0	0
11	.14	0	0
12	.13	0	0

```
                    % OF NITRATE PRESENT
                      10   20   30   40   50   60
          LEVEL 1    ***********************
          LEVEL 2    *********
          LEVEL 3    ****
          LEVEL 4
          LEVEL 5
          LEVEL 6
          LEVEL 7
          LEVEL 8
          LEVEL 9
          LEVEL 10
          LEVEL 11
          LEVEL 12
```

TIME AFTER TREATMENT = 20 DAYS

STRATA NO.	MEAN MOISTURE CONTENT	MEAN NITRATE CONTENT	% NITRATE CONCENTRATION
1	.02	8.45616	4.50456
2	.08	2.10349	1.12052
3	.155	5.61136	2.98915
4	.232	12.1055	6.44855
5	.27	18.6343	9.92638
6	.29	23.6898	12.6195
7	.28	24.4283	13.0129
8	.28	24.055	12.814
9	.28	22.5948	12.0362
10	.25	18.4927	9.85097
11	.23	15.2561	8.12687
12	.21	12.297	6.55055

```
                    % OF NITRATE PRESENT
                    10  20  30  40  50  60
         LEVEL 1
         LEVEL 2
         LEVEL 3
         LEVEL 4    **
         LEVEL 5    ***
         LEVEL 6    *****
         LEVEL 7    *****
         LEVEL 8    *****
         LEVEL 9    ****
         LEVEL 10   ***
         LEVEL 11   ***
         LEVEL 12   **
```

TIME AFTER TREATMENT = 40 DAYS

STRATA NO.	MEAN MOISTURE CONTENT	MEAN NITRATE CONTENT	% NITRATE CONCENTRATION
1	.24	.162048	.410731
2	.24	.454121	1.15103
3	.25	.82444	2.08965
4	.26	1.2214	3.09579
5	.27	1.74158	4.4125
6	.29	2.7001	6.84375
7	.28	3.59552	9.11331
8	.28	4.6951	11.9003
9	.28	5.80105	14.7035
10	.25	6.01056	15.2345
11	.23	6.16274	15.6203
12	.21	6.08486	15.4229

```
                   % OF NITRATE PRESENT
                   10  20  30  40  50  60
         LEVEL  1
         LEVEL  2
         LEVEL  3
         LEVEL  4
         LEVEL  5
         LEVEL  6    **
         LEVEL  7    ***
         LEVEL  8    ****
         LEVEL  9    *****
         LEVEL 10    ******
         LEVEL 11    ******
         LEVEL 12    ******
```

REFERENCES

1. W.R. Gardner and R.H. Brooks 'A descriptive theory of leaching', *Soil Science* **83**, 295–304 (1957)
2. W.R. Gardner 'Movement of Nitrogen in Soil', *Soil Nitrogen* (eds W.V. Bartholomew and F.E. Clark), No. 10 in Agronomy series (Madison American Society of Agronomy, 1965)
3. A.E. Sheidegger *The Physics of Flow through Porous Media. Part X. Mixible Displacement* (Toronto, University of Toronto Press, 1960) pp. 256–60
4. I.G. Burns *Journal of Soil Science* **25**, 165–178 (1974)
5. I.G. Burns *Journal of Agricultural Science, Cambridge* **85**, 443–454 (1975)
6. R.W. Terkeltoub and K.L. Babcock *Soil Science* **111**, 182–187 (1971)
7. T.M. Addiscott and D. Cox *Journal of Agricultural Sciences, Cambridge* **87**, 381–389 (1976)
8. T.M. Addiscott *Journal of Soil Science* **28**, 554–563 (1977)
9. T.M. Addiscott *et al., Journal of Soil Science* **29**, 305–314 (1978)
10. A. Barnes *et al., Journal of Agricultural Science, Cambridge* **86**, 225–244 (1976)

5.

Describing Plant Growth

N.C. Steele

5.1 STATEMENT OF THE PROBLEM

What is growth? How do the various parts of a plant grow in relation to one another? The first question is apparently easily answered — at least until one thinks about the problem for a while. The second question is not just one of academic interest: efficient agriculture requires that, for example, root crops should yield the maximum amount of edible material with the minimum waste. However, is there a maximum or minimum? Clearly, the two questions are inextricably related, and an understanding of the phenomenon would be economically valuable, particularly if such understanding indicates methods of producing 'better' varieties of root vegetables. Problems of this magnitude are obviously beyond our scope. However, it is perhaps possible to consider the basic phenomenon of plant growth in some way. Specifically we will try to construct a model of plant growth capable of describing the *relative* development of plant parts.

5.2 INITIAL STUDENT REACTION

At this stage time should be allowed for thought. Growth will be defined as an increase in 'size', in weight or perhaps in volume. How does one measure these? Perhaps the idea of dry weight will be mentioned, with growth defined as an increase in dry weight over a time interval. In discussing plant parts, it will soon be clear that further definitions are required — what is a root for example? Finally, in trying to describe the growth process it will be clear that some knowledge of the internal mechanisms of a plant is needed.

5.3 BACKGROUND TO THE PROBLEM

To construct a model of plant growth some rules must be laid down. A useful concept is that of the instantaneous *state* of the plant defined by the instantaneous values of some *properties* of the plant (or plant part). A property of the plant changes by means of a *process* and these processes occur at rates which are determined by the state of the plant system.

Table 5.1, due to Thornley[1], lists some of the important plant processes and properties.

Table 5.1 Plant Processes and Properties

Processes	Properties
Light absorption, reflection scattering, transmission	Size, weight, number, length area, volume
Nutrient uptake Water uptake	Structure, internal structure, external form, geometry, position, pattern
Within-plant transport of gases, sugars, hormones, growth factors, other nutrients	Composition, chemical concentrations
Photosynthesis	Temperature
Respiration maintenance respiration	Pressure
Utilization of carbon assimilates, other nutrients	Electrical and optical properties absorptivity, conductivity, reflectivity, scattering, transmissivity
Other aspects of metabolism	Mechanical properties
Wastage, loss of material	elasticity, permeability
Storage	
Growth	
Development germination morphogenesis reproduction senescence	
Harvesting	
Post-harvest processes	

Table © by Academic Press Inc. (London) Ltd.

Obviously, any attempt to include most of these processes and properties into a simple model would be doomed to failure, and perhaps guided by the biologists[2] we must try to isolate the dominant processes for at least part of the growth period. Let us look at some possibilities in detail.

Most people have heard of photosynthesis; this is the process of manufacture of nutrients, or substrates, in the form of carbon by the leaves of a plant. The process takes its name from the requirement of light for the 'fixing' reaction to take place. Nutrients are also taken up from the soil through the root system and here the substrate is nitrogen. Thus we can identify two processes by which a plant can gain material, but are there any processes which cause loss of material? The answer is, of course, yes!

An obvious, but easily overlooked, method is by the death of leaves. Less obvious is the process of maintenance respiration, by which the plant removes the waste products generated in the manufacture and conversion of nutrients. Clearly, then, the growth process can be thought of in terms of a change of the plant state, reflecting a balance between processes which cause an increase in plant dry weight and those which cause a loss. Actual growth results provided the positive contributions exceed the negative contributions. Notice that here we have isolated the property of the plant that will be of final interest to us — namely, the plant part dry weight.

Having identified some processes and an overall property to be used in our model, we must now decide upon a suitable division of a plant into constituent parts. If we are considering root crops (potatoes, turnips, etc.), a reasonable division would be into (a) fibrous roots, (b) storage organ (i.e. usable crop), and (c) stem/leaves. Notice that (a) and (c) are production areas, while (b) represents the harvestable quantity. Also at this stage assumptions must be made about the internal mechanisms of the plant for the distribution of nutrient; in other words, the mechanism by which the process rates will depend on the state of the plant. A definite, if arbitrary, choice is needed, since these mechanisms can not be regarded as 'known' in any real sense. It is known, however, that carbon is transported throughout the plant by the 'phloem' system and nitrogen by the 'xylem' system. Many models are possible[1] for the working of these systems, specifically to explain how substrate produced in one plant part is made available to the other parts. One approach is based on the concept of a circulatory system which assumes that nutrients once absorbed by (or manufactured within) the plant are then generally available to all parts according to demand.[2] It must be emphasised that to select this model is an *arbitrary* choice and the alternative models[1] should not be ignored.

5.4 STUDENT REACTION

Student reaction will probably be fairly harsh in view of the lack of a 'proper' model. It may well be worth explaining that established models, for example those of Newtonian mechanics, stand or fall on their ability to predict experimental results, and in this respect, any model selected here is no different. It would be appropriate here to draw attention to the essential idea of the *balance* between factors leading to a dry weight *gain* and those leading to a dry weight *loss*, suggesting the idea of an *equation* to represent this balance. With this in mind, appreciation of time dependency is essential, leading to the proposal of a differential (or perhaps a difference) equation model. The final task at this stage of the formulation is to encourage thought about how the various parts of the plant should be treated and, especially, how the gain/loss processes can be represented within the framework established.

5.5   AN AGREED APPROACH TO THE PROBLEM

Let us summarise thinking so far. The plant is to be split into three parts: fibrous roots, storage organ and shoots/leaves. Nutrient supply is to be by a circulatory system, with each plant part able to draw from the total supply according to its demand. Loss of dry weight will be by leaf death and by respiration.

Now, some progress is possible: it is reasonable to try to write equations for the balance between dry weight increase and loss for each part of the plant. If a differential equation model is sought, then attention is focussed on events in a time interval δt. Turning to the biological processes, a form has to be invented for nutrient assimilation and for respiration.

For nutrient assimilation, we have assumed that each part will make some demand on the total supply. What will determine this demand? Recall that our overall assumption is that the rate of the process (of assimilation) will depend on the current state, as measured by a property or properties of the plant part at a particular instant. What property should we use? A reasonable choice is the size (dry weight) of an individual component, so that in a time interval each component will absorb a quantity of the available nutrient proportional to its current size, measured as a dry weight.

The process of respiration applies to all components in the sense that waste products are removed from all parts of a plant. The process rate here is assumed to be determined by two properties, the amount of nutrient assimilated (digestion!) and by the size of the plant part. Thus in a time interval each component will remove matter (measured as a dry weight) in two ways: an amount proportional to the amount of nutrient absorbed in that interval, and an amount proportional to its current size.

The final process to be included is loss of matter due to leaf death. This is of course confined to one component only, and since there is no real guidance as to which property or properties this process rate is likely to depend upon, we will defer a decision until one is forced upon us.

5.6   A SOLUTION TO THE PROBLEM

Fig. 5.1   Events in time interval δt

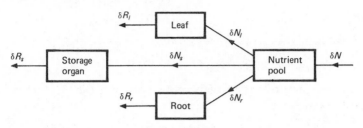

Fig. 5.1 illustrates events in a time interval $(t, t + \delta t)$, with a total amount δN of nutrient being available within this time interval. Assuming that nutrient is conserved, and that it is all taken up, we have

$$\delta N = \delta N_l + \delta N_s + \delta N_r$$

where $\delta N_l, \delta N_s, \delta N_r$ are the amounts of nutrient supplied to the leaves, storage organ and roots respectively, during the time interval $(t, t + \delta t)$. Utilising the demand concept we can write

$$\delta N_l = A_l l \delta N$$
$$\delta N_s = A_s s \delta N \qquad\qquad [5.1]$$
$$\delta N_r = A_r r \delta N$$

where l, s and r are the instantaneous dry weights of leaves, storage organ and roots respectively. Here, A_l, A_s and A_r can be thought of as relative demand strengths for the various organs. Notice that

$$\delta N = (A_l l + A_s s + A_r r)\,\delta N$$

i.e. $A_l l + A_s s + A_r r = 1$

so that the As cannot be constant, since l, s and r depend on t.

Consider now the growth process during the time interval δt. Assuming that for the whole plant the increase in dry weight in this interval is δW, then

$$\delta W = \delta N - \delta R - \delta D_l$$

where δR is the total dry weight loss due to respiration and δD_l, the loss of dry weight due to leaf death. Clearly $\delta W = \delta l + \delta s + \delta r$,   the sum of the dry weight increases for each component, but

$$\delta l = \delta N_l - \delta R_l - \delta D_l$$
$$\delta s = \delta N_s - \delta R_s \qquad\qquad [5.2]$$
$$\delta r = \delta N_r - \delta R_r$$

where δR_l, δR_s and δR_r are the *partial* respiratory dry weight losses. An appropriate form must be used to represent these respiratory losses, and as discussed earlier we write

$$\delta R_l = \alpha_l \delta N_l + m_l l \, \delta t$$

$$\delta R_s = \alpha_s \delta N_s + m_s s \, \delta t \qquad\qquad [5.3]$$

$$\delta R_r = \alpha_r \delta N_r + m_r r \, \delta t$$

where the αs may or may not be constants and the ms are respiration rates/unit mass.

Substituting [5.3] into the fundamental balance Equation [5.2] and letting $\delta t \to 0$, we obtain

$$\frac{dl}{dt} = \beta_l \frac{dN_l}{dt} - m_l l - \frac{dD_l}{dt}$$

$$\frac{ds}{dt} = \beta_s \frac{dN_s}{dt} - m_s s$$

$$\frac{dr}{dt} = \beta_r \frac{dN_r}{dt} - m_r r$$

where $\beta_l = 1 - \alpha_l$, etc.

These three equations give the rate of increase in plant part dry weight in terms of the total available nutrient; however, using the demand relationships [5.1] as $\delta t \to 0$, we can write

$$\frac{1}{A_l l} \frac{dN_l}{dt} = \frac{1}{A_s s} \frac{dN_s}{dt} = \frac{1}{A_r r} \frac{dN_r}{dt} = \frac{dN}{dt} \qquad\qquad [5.4]$$

Thus we can eliminate the nutrient supply rate terms to determine three equations for the relative growth rates of the three plant parts as

$$\frac{1}{\beta_l A_l l} \left(\frac{dl}{dt} + m_l l + \frac{dD_l}{dt} \right) = \frac{1}{\beta_s A_s s} \left(\frac{ds}{dt} + m_s s \right) = \frac{1}{\beta_r A_r r} \left(\frac{dr}{dt} + m_r r \right)$$

At this stage we have solved the original problem in the sense that given growth data for any one plant part, then the growth of the other and indeed the plant as a whole can be predicted.

5.7 A PARTICULAR PLANT PART RELATIONSHIP

As a prelude to model validation it is useful to consider in more detail the relationship between leaf dry weight and storage organ dry weight. Is it possible to obtain an algebraic form capable of being checked against experimental data? In effect this means that the differential equation

$$\frac{1}{\beta_l A_l l} \left(\frac{\mathrm{d}l}{\mathrm{d}t} + m_l l + \frac{\mathrm{d}D_l}{\mathrm{d}t} \right) = \frac{1}{\beta_s A_s s} \left(\frac{\mathrm{d}s}{\mathrm{d}t} + m_s s \right)$$

has to be integrated. Clearly some assumptions have to be made concerning the As, the βs and the ms. In the first place we know that the As cannot be constant, but it is possible that $A_l/A_s = \mu$ is constant, meaning that the ratio of the specific rates of assimilation is constant. If we can assume the βs and ms are also constant, then the only obstacle to integration is the leaf death term $\mathrm{d}D_l/\mathrm{d}t$.

Progress is possible, however, if we force this term into an amenable form: write

$$\frac{\mathrm{d}D_l}{\mathrm{d}t} = d_l l$$

making d_l an *average* death rate for leaves (grams per gram per day perhaps). Notice that d_l *is* a measurable quantity — an important consideration!

Making all these assumptions we have

$$\left(\frac{1}{l} \frac{\mathrm{d}l}{\mathrm{d}t} + m_l + d_l \right) = \frac{\mu \beta_l}{\beta_s} \left(\frac{1}{s} \frac{\mathrm{d}s}{\mathrm{d}t} + m_s \right)$$

and integrating from t_0 to t gives

$$\ln l - \ln l_0 = \gamma(\ln s - \ln s_0) + (\gamma m_s - m_l - d_l)(t - t_0), \qquad (\gamma = \mu \beta_l / \beta_s).$$

or $\quad \ln l = a + \gamma \ln s - bt$

with

$$a = \ln l_0 - \gamma \ln s_0 - t_0 (\gamma m_s - m_l - d_l) = \ln l_0 - \gamma \ln s_0 + bt_0$$

where

$$b = (-\gamma m_s + m_l + d_l).$$

So finally we have a relationship between the dry weights of leaves and storage organ in the form

$$\ln l = a + \gamma \ln s - bt \qquad\qquad\qquad [5.5]$$

with a, b and γ constant.

5.8 MODEL VALIDATION

Table 5.2, which is derived from Barnes,[2] data for harvests of carrots in 1978, gives the relationship between ln(dry leaf weight) and ln (dry storage organ weight) for various times after emergence. If our model is valid then it should be possible to obtain values of the parameters a, γ and b so that the data of Table 5.2 satisfy Equation [5.5]. This confirmation is left as an exercise for the student.

Table 5.2 Grouped carrot data

Time from emergence (days)									
35		63		91		119		140	
$\ln s$	$\ln l$	$\ln s$	$\ln l$	$\ln s$	$\ln l$	$\ln s$	$\ln l$	$\ln s$	$\ln l$
-3.8	-2.5	-1.9	-1.3	-1.7	-1.4	-1.3	-1.4	-0.6	-1.2
-1.9	-1.0	-0.2	-0.1	0.7	0.4	1.0	0.1	1.3	0.3
-1.1	-0.5	0.6	0.6	1.4	0.9	1.7	0.8	2.0	0.8
-0.4	0.2	1.2	1.1	2.3	1.6	2.7	1.6	2.8	1.4

5.9 FURTHER WORK

1. Assume that for short periods of time, the growth of the storage organ is exponential, so that $s = s_0\, e^{q(t-t_0)}$, where q is a growth rate parameter. Does this enable you to obtain a simpler form of Equation [5.5]?

2. Over longer periods of time growth is restricted and an overall maximum size, s_∞ say, is approached by the storage organ. The differential equation describing the growth process in [5.1] is

$$\frac{ds}{dt} = qs \quad \text{with} \quad s(t_0) = s_0$$

can you propose a new equation modelling this new form of restricted growth? Can you solve it? What does Equation [5.5] become in this case?

REFERENCES

1. J.H.M. Thornley *Mathematical Models in Plant Physiology*, (London, Academic Press, 1976)
2. A. Barnes 'Vegetable plant part relationships. II. A quantitative hypothesis for shoot/storage root development', *Ann. Bot.*, **43**, 487–499 (1979)

6.

Tumour Growth and the Response of Cells to Irradiation

J.R. Usher and D.A. Abercrombie

6.1 INTRODUCTION TO THE LECTURER

Many fundamental contributions to research into the treatment of cancer have been made by workers who have drawn from such disciplines as medicine, biology, biochemistry, physiology and physics. Indeed much research is difficult to categorise and is decidedly interdisciplinary in nature. Mathematics has also played an incisive role, and there is evidence that in cancer research, as in many other fields of human endeavour, its role is becoming increasingly more important.

The present authors have developed a group of related case studies[15] based on their own work and on the work of others. The intention of that development was to demonstrate, in a form suitable for classroom use, one of the most important methodological approaches to the development of a mathematical model of some 'real life' phenomenon, viz. the breaking down of a problem into identifiable (and tractable) interrelated parts, examining each component separately and reconstituting the parts into a coherent whole. The intention of the work in the following pages is to present a development of two particular components of the above study in more detail with the intention that the participation in such a development will gain for the student some experience in and some insight into the general modelling process.

6.2 PRIMARY PROBLEMS AND BACKGROUND MATERIAL

6.2.1 STATEMENT OF THE PRIMARY PROBLEMS

There are three main modes of treatment of cancer: surgery, radiotherapy and chemotherapy. While a study of each of the latter two modes can be approached from a mathematical viewpoint it is treatment by radiotherapy to which the following studies are ultimately directed.

Obviously the therapists' declared aim is to administer the 'best possible' treatment. Just what this means is discussed in References 1 and 15. However, before detailed consideration can be given to optimal treatment schedules, two questions (but not the only two!) to be answered are:

(a) How do tumours grow?

(b) What is the effect of irradiation on tumour cells?

6.2.2 GENERAL BACKGROUND MATERIAL

Most tumours exhibit the following characteristics:

(a) growth is relatively free from the body's homeostatic control mechanisms (i.e. those mechanisms which attempt to maintain equilibrium between the body and its environment);

(b) power of invasion and replacement of surrounding normal tissue; and

(c) potential facility to metastasize (i.e. to form secondary tumours at a remote site).

Both normal and tumour cells are derived by mitosis. This is the process of cell division: each chromosome splits into two, one of the resulting duplicates passing to each of the daughter cells. In the case of normal cells they develop into different classes each with a specific role to play, e.g. liver cells, and this process is referred to as 'differentiation'.

In treating a tumour by irradiation the tumour is bombarded by ionising radiation with the specific purpose of inflicting maximum damage on the tumour while keeping damage to surrounding and enclosed normal tissue below that level above which the normal tissue cannot recover. It has been found that normal tissue has a greater ability than tumour cells to recover following a fractionated (i.e. split) dose.[2]

6.3 TUMOUR GROWTH

6.3.1 BACKGROUND MATERIAL

For ethical reasons estimation of tumour growth is difficult. Resort has to be made to available literature on the modelling of biological populations.[15]

Knowledge of tumour growth is relatively scant but is summarized in Section 4 of Reference 15. Attention will be focused on the following points.

1. Tumours normally become clinically observable when the total number of tumour cells exceeds 10^{11} or thereabout.

2. When first detected, cells in many tumours appear to be doubling in number over a constant period of time (this time period is often referred to as the doubling time).

3. At a later stage of evolution competition for limited space tends to produce a stable asymptotic population size.

6.3.2 CASE STUDY 1

Develop a model to describe tumour growth.

6.3.3 A SOLUTION

The characteristics of tumour growth noted in Section 6.3.1 may be incorporated into a two-stage model as follows.

(a) Develop a model to describe tumour growth when the tumour is first detected.

(b) Develop a model to describe the situation when tumour cells are competing for limited space.

Case (i)

Since the rate of growth of a tumour will appear to an observer to increase continuously rather than by sudden jumps a continuous model would seem appropriate. An attempt will now be made to develop a continuous model which describes the situation when the tumour cells are doubling in number over a constant period of time.

It is assumed that the increase in the number of tumour cells in a small time interval $(t, t + \Delta t)$ is proportional to Δt and to the population size at time t. If $m(t)$ denotes the number of cells at time t, then

$$m(t + \Delta t) - m(t) = \lambda m(t)\, \Delta t \qquad \qquad [6.1]$$

λ being the constant of proportionality.

In the limit as $\Delta t \rightarrow 0$ the relative (or specific) growth rate $(1/m)(\mathrm{d}m/\mathrm{d}t)$ is seen to be constant, i.e.

$$\frac{1}{m}\frac{\mathrm{d}m}{\mathrm{d}t} = \lambda. \qquad \qquad [6.2]$$

(It is worth noting here that in most problems concerning biological growth the relative rate of growth is fundamentally more important than the absolute rate of growth.)

A suitable solution of differential equation [6.2] is easily seen to be given by

$$m(t) = m(0)\exp \lambda t \qquad \qquad [6.3]$$

i.e. the tumour cells are proliferating exponentially with relative growth constant λ. Since this model is intended to describe the growth behaviour of a tumour when the tumour first becomes clinically observable it may be assumed that

$$m(0) = 10^{11}. \qquad \qquad [6.4]$$

It is suggested that the student is left the task of showing how the constant λ is related to a quantity possessing physical significance, namely

$$\lambda = \frac{\ln 2}{\tau} \qquad \qquad [6.5]$$

where τ denotes the doubling time, since a very useful function of the modelling process is to relate mathematical constants with identifiable physical components.[4] In this task the student might find the following functional relation useful:

$$m(t + \tau) = 2m(t). \qquad \qquad [6.6]$$

An exponential growth curve is illustrated in Fig. 6.1.

Clearly, the continuous model just developed may be valid initially but it will be unsuitable once the tumour cells compete for limited space.

Case (ii)

In the case where tumour cells compete for limited space and eventually the population size tends to stabilize it seems reasonable to assume that the relative rate of growth is no longer constant, as in case (i), but a function of the number of tumour cells present, i.e.

Fig. 6.1 Exponential growth curve

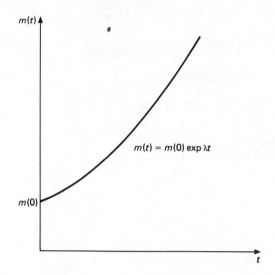

$$\frac{1}{m}\frac{dm}{dt} = \mu(m)$$ [6.7]

where $m(t)$ denotes the number of tumour cells at time t and $\mu(m)$ denotes some decreasing function of m.

A suitable expression for $\mu(m)$ must now be found. A linear expression would appear to be the simplest, i.e.

$$\mu(m) = A + Bm$$ [6.8]

where A and B are constants.

Again, as in case (i), it is suggested that the student is left the task of showing how these constants are related to quantities possessing a physical significance. The appropriate expression which decreases from the value λ (as given in case (i)) to the value 0 is given by

$$\mu(m) = \frac{\lambda}{\theta - m(0)}(\theta - m)$$

where the parameter θ is the asymptotic population size. It is hoped that the two parameters λ and θ may be determined from the growth characteristics of the tumour under investigation. This formulation gives rise to a growth rate governed by the differential equation

$$\frac{dm}{dt} = \frac{\lambda m}{\theta - m(0)}(\theta - m).$$ [6.9]

Such a rate of growth is often referred to as a 'Verhulst' growth rate.[16]

By separating the variables in differential equation [6.9] and integrating, after some algebra, the following solution is obtained:

$$m(t) = m(0) \left[\frac{m(0)}{\theta} + \left(1 - \frac{m(0)}{\theta}\right) \exp\left(\frac{-\lambda\theta t}{\theta - m(0)}\right) \right]^{-1}. \qquad [6.10]$$

A Verhulst growth curve is illustrated in Fig. 6.2.

Other suitable expressions for $\mu(m)$ in Equation [6.7] may be investigated.

Fig. 6.2 Verhulst growth curve

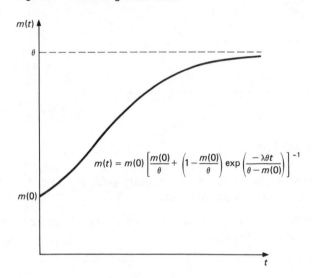

6.3.4 FURTHER DISCUSSION

The two-stage model discussed in Section 6.3.3 provides a starting point for developing more realistic models. In practice it has been found that the exponential and Verhulst growth curves are often not adequate for fitting to experimental data. There is experimental evidence to suggest that many tumours appear to exhibit Gompertzian-type growth behaviour[3,12,13]. Such behaviour is governed by the differential equation

$$\frac{dm}{dt} = -\lambda m \ln\left(\frac{m}{\theta}\right) \qquad [6.11]$$

where λ, $\theta \, (> 0)$ are parameters that may be determined from the growth characteristics of the tumour under investigation.

A solution of differential equation [3.11] may be written in the form

$$m(t) = m(0) \exp \left\{ \left[\ln \left(\frac{\theta}{m(0)} \right) \right] [1 - \exp(-\lambda t)] \right\}. \qquad [6.12]$$

A representative Gompertz growth curve is shown in Fig. 6.3 (see curve A) where a semi-logarithmic scale has been employed to facilitate comparison with representative exponential and Verhulst growth curves.

One of the present authors, J.R. Usher[14] has developed a more general tumour growth model which incorporates the exponential, Verhulst and Gompertzian growth behaviour as special cases. In this work the differential equation governing possible tumour growth is given by

$$\frac{dm}{dt} = \frac{\lambda m}{\alpha} \left[1 - \left(\frac{m}{\theta} \right)^{\alpha} \right] \qquad [6.13]$$

where the parameters $\alpha (\geqslant 0)$, λ, $\theta (> 0)$ are again determined from the growth characteristics of the tumour under investigation.

Fig. 6.3   Typical growth curves: A, Gompertz (Equation [6.12]); B, Verhulst; C, exponential, $\lambda = 0.01$, $m(0) = 10^{11}$, $\theta = m(0) \exp K$, $K = 28.5$

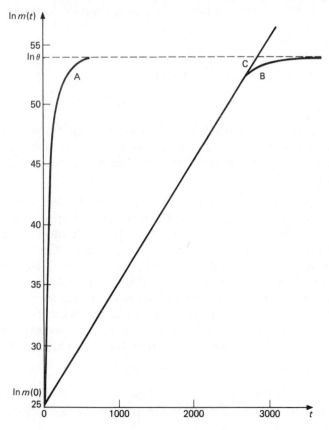

A solution to this differential equation may be written in the form

$$m(t) = m(0) \left\{ \left(\frac{m(0)}{\theta} \right)^{\alpha} + \exp(-\lambda t) \left[1 - \left(\frac{m(0)}{\theta} \right)^{\alpha} \right] \right\}^{-1/\alpha} \qquad [6.14]$$

The study of this general model automatically incorporates the following limiting forms:

(a) $\alpha \to 0$: $\qquad \dfrac{dm}{dt} = -\lambda m \ln \left(\dfrac{m}{\theta} \right) \qquad\qquad [6.15]$

i.e. Gompertzian growth rate

(b) $\alpha \to 1$: $\qquad \dfrac{dm}{dt} = \lambda m - \dfrac{\lambda}{\theta} m^2 \qquad\qquad [6.16]$

i.e. Verhulst growth rate

(c) $\alpha \to 1$, $\qquad \dfrac{dm}{dt} = \lambda m \qquad\qquad\qquad [6.17]$
$\quad\;\; \theta \to +\infty$:

i.e. exponential growth rate

(N.B. Care must be shown when taking such limiting forms to ensure their existence.)

6.4 RESPONSE OF CELLS TO IRRADIATION

6.4.1 BACKGROUND MATERIAL

In the previous section we dealt with models of tumour growth in which we derived mathematical descriptions of natural tumour development. In this section we concern ourselves with the modelling of the effect of a single radiation insult.

It is known that when a population of cells is bombarded with ionising radiation, a proportion of the population is killed. Further, the greater the dose, the smaller is the proportion of cells which survive the insult. In 1956, Puck and Marcus[11] described the quantitative relationship between surviving fraction of Hela S3 tumour cells and X-ray dose. The graphical representation of this relationship is reproduced in Fig. 6.4. Such curves are conventionally plotted as log (surviving fraction) against dose and are called 'survival' curves. Curves of this form had been observed previously for the radiation killing of bacteria. However, Puck and Marcus were the first to observe such a curve for mammalian cells.

The main characteristics of survival curves are:

(a) they are asymptotic to a straight line for large dosages; the point where this asymptotic line cuts the log (surviving fraction) axis is called the extrapolation number and may vary over a large range of values for different cell lines and types of radiation;

(b)   they exhibit a 'shoulder' at relatively small dosages.

Fig. 6.4   Survival of reproductive capacity in Hella cells as a function of X-ray dose[11]

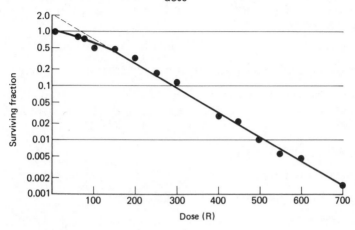

Dose (R)

6.4.2   CASE STUDY 2

Derive a mathematical model to 'explain' the shape of the survival curve.

6.4.3   A SOLUTION

It will be apparent that the solution presented here is just one of a number of possibilities and in fact it is unlikely that the data available would be sufficient to discriminate between two (or more) models which 'explain' the shape of the survival curve. It is quite possible that enthusiastic students will produce quite ingenious models. However, if it is desired to direct the students towards the 'classical' solution, the following ideas should be suggested gradually to the students.

1.   The shoulder on the survival curve illustrates that there is an accumulation of damage before the cell is killed. The second fifty rads kills a greater proportion of cells than the first.

2.   X-rays can be regarded as 'packets' of energy with which we bombard the cells. Thus irradiation can be likened to throwing ping-pong balls at a collection of buckets (a visual model!). It may take several ping-pong balls to kill a bucket.

Such considerations may lead to the following solution.

Assume:

(a) each cell has n 'targets'; the necessary and sufficient condition for a cell to die is defined to be that each of the n targets receives at least one 'hit' due to the radiation;

(b) the number of hits received by a randomly chosen target is a Poisson variate with a mean proportional to the total dose of radiation;

(c) each target receives hits independently of all others.

Encourage the students to discuss the weaknesses of this model. |

The assumptions outlined above lead to the classical multi-target model widely used by radiobiologists. The analysis for this model is now presented.

From assumption (b) above

P (a given target receives no hits)* $= \exp(-kD)$

where D denotes the dose and k is a constant.

∴ P (a given target receives at least one hit) $= 1 - \exp(-kD)$.

From assumption (c)

P (all n targets receive at least one hit) $= [1 - \exp(-kD)]^n$.

From assumption (a)

P (a given cell survives) $= 1 - [1 - \exp(-kD)]^n$.

The constant k is commonly written as $1/D_0$, where D_0 is the dose required to produce an average of one hit per target and D_0 is known as the 'mean lethal dose'.

The expected proportion of survivors following a dose D, more commonly referred to as the *clonogenic surviving fraction* and denoted by σ, is given by

$$\sigma = 1 - \left[1 - \exp\left(-\frac{D}{D_0}\right)\right]^n.$$ [6.18]

Now consider the behaviour of the associated survival curve. For large D/D_0:

$$\sigma \sim n \exp\left(-\frac{D}{D_0}\right)$$

$$\ln \sigma \sim \ln n - \frac{D}{D_0}.$$

*$P(x)$ denotes the probability that event x will occur.

Thus a semi-log plot of σ versus D approaches a straight line with slope $-1/D_0$ and with zero dose intercept $\sigma = n$ (this is the reason why n is referred to as the extrapolation number).

When $n = 1$,

$$\ln \sigma = - \frac{D}{D_0}$$

The associated survival curve has no shoulder. Cells with such a survival curve are said to undergo exponential survival.

When $n > 1$ the survival curve has a shoulder which becomes more distinct with increasing n.

Puck and Marcus[11] found that equation [6.18] fitted their data 'well' when $n = 2$, and $D_0 = 96$ rad (see Fig. 6.4).

Note 1

We must avoid the conclusion that there are two 'targets' in Hela S3 cells *since the validity of the mechanism behind a model is 'not' proved just because the model fits.*

Note 2

The parameters n and D_0 are related to the radiosensitivity of the cell to the particular type of radiation being employed. There is certainly no guarantee that different cell types subjected to different kinds of radiation would yield similar values.

6.4.4 CRITICISMS OF THE MODEL

1. For a given set of experimental data the extrapolation number n will in general be non-integer and therefore cannot represent 'exactly' the number of targets in a cell, but obviously could represent some type of 'average'.

2. For $n > 1$ the survival curve associated with Equation [6.18] has zero gradient when $D = 0$, which is not the case for all available experimental data.

3. The assumption that *each* cell receives *at least one* hit is somewhat arbitrary and obviously many similar combinations are possible. These other possibilities might be examined.

4. The model takes no account of repair processes which are known (by experimentation) to be operating.

6.4.5 FURTHER DISCUSSION

A Repair Mechanism

Laurie, Orr and Foster[9] proposed a model which specifically incorporated a repair mechanism. This model, which was of a coupled differential equation type, suggests the following more naive probabilistic model.

Powers[10] postulated the existence of a pool of intracellular constituents which could be used up in repairing potentially lethal damage in a cell. Thus, suppose that on irradiation a cell possesses a random number of sites of potentially lethal damage. Immediately the intracellular repair processes are activated each site of damage is repaired with a probability which is dependent on the total dose administered. Sites not repaired become sites of developed lethal damage.

Exercise

Make appropriate assumptions and construct a model which 'explains' the shape of a typical survival curve.

Solution

Let X be the number of sites of potentially lethal damage before repair, and let Y be the number of sites of developed lethal damage after the repair processes have been activated.

It is reasonable to assume that X has a Poisson distribution with mean proportional to the total dose D administered, i.e.

$$P(X = r) = \frac{e^{-kD}(kD)^r}{r!}, \quad r = 0, 1, 2, \ldots$$

(k being a constant of proportionality).

Suppose that a given cell has r sites of potentially lethal damage before repair then the probability that there are s sites of developed lethal damage after repair $(s \leqslant r)$ is precisely the probability that $r-s$ sites of potentially lethal damage are repaired. If each site is repaired (or not) independently of all others and with probability π, where π is dependent on the total dose D administered, then we have

$$P(Y = s \mid X = r) = \binom{r}{r-s} \pi^{r-s}(1-\pi)^s, \quad 0 \leqslant s \leqslant r$$

and

$$P(Y = s) = \sum_{r=s}^{\infty} \binom{r}{r-s} \pi^{r-s}(1-\pi)^s \frac{e^{-kD}(kD)^r}{r!}, \quad s = 0, 1, 2, \ldots$$

$$= \frac{[(1-\pi)kD]^s\,e^{-(1-\pi)kD}}{s!}, \qquad\qquad s = 0, 1, 2, \ldots$$

i.e. the number of sites of developed lethal damage is a Poisson variate with mean $(1-\pi)kD$.

Now if we assume that a cell is viable only if it contains no sites of developed lethal damage, then we have

$$\sigma = P(\text{survival}) = P(Y = 0)$$
$$= e^{-(1-\pi)kD}$$

and thus

$$\ln \sigma = -(1-\pi)kD.$$

If it is recalled π is dependent on D, then we need a suitable functional form for $\pi(D)$ which gives rise to a survival curve with the required characteristics. It is now shown that the following functional form meets these requirements:

$$\pi(D) = \frac{1}{1+aD}.$$

Note that this form satisfies the conditions $\pi(0) = 1$ and $\pi(\infty) = 0$ — both realistic requirements.

Now

$$\ln \sigma = \frac{-akD^2}{1+aD}$$

$$= -kD + \frac{k}{a} - \frac{k}{a+a^2D}. \qquad\qquad [6.19]$$

The survival curve with equation given by [6.19] is thus asymptotic to a straight line with slope $-k$ and the extrapolation number is $\exp(k/a)$.

Notes

1. The form $\pi(D) = 1/(1+aD)$ was chosen to ensure that the survival curve possessed the required characteristics. Thus no particular radiobiological intuition is claimed in the proposal of this form. However, it can be seen that for a fixed dosage, as a increases, the probability of repair decreases and if also the slope of the asymptotic line is fixed the extrapolation number decreases, both of which signify a radiosensitive cell line.

2. The survival curve associated with Equation [6.19] is horizontal at $D = 0$ and would be unsuitable to represent those empirical curves which suggest a negative slope at the origin.

6.4.6 FURTHER EXERCISES

1. Fit a curve with an equation of the form [6.19] to the data of Puck and Marcus (see Fig. 6.4) by equating extrapolation numbers and asymptotic slopes.

2. Suggest other functional forms for $\pi(D)$ which are consistent with the shape of a typical survival curve.

6.5 SUMMARY AND CONCLUSION

The cancer therapist's ultimate aim in the treatment of cancer is the total eradication of all tumour tissues. If treatment is to be by radiotherapy, then an alternative to evolving the necessary optimal treatment schedules by clinical trial and error is to construct a mathematical model of the radiobiological and cellular processes involved. In the preceding pages two components of these complex processes have been isolated and, based on a little knowledge of the biological processes and on empirical observation, elementary models have been formulated and discussed. The two components discussed were the dynamics of tumour growth in the absence of irradiation and the response of tumour cells to irradiation.

One further component of fundamental importance to the problem is the response of normal tissue to irradiation. It is the effects of irradiation on the normal tissue throughout and surrounding the tumour volume which prevent the radiotherapist from bombarding the tumour with arbitrarily large doses. The normal tissue, whilst receiving some injury, must not be damaged beyond the level above which it cannot recover. Ellis[5-7] and Kirk *et al.*[8] have developed empirical models which describe the response of normal tissue to irradiation. Abercrombie[1] has criticised these models and indicated a possible alternative formulation which takes more account of the biological processes known to be operating.

The synthesis of these three components permits the development of a model for determining optimal radiotherapy treatment schedules. Wheldon and Kirk[17] derived optimal schedules in the case of exponential tumour growth, and Usher[14] extended the analysis to tumours whose associated growth curves belong to a large class of curves which include the exponential, the Gompertzian and the Verhulst growth curves as special cases. It is significant that classical differential calculus provided sufficient tools for these investigations. Usher and Abercrombie[15] discuss the entire problem in the form of a sequence of case studies suitable for classroom presentation.

The development of models for the investigation of the treatment of cancer by radiotherapy is an active research area. It is hoped that the preceding pages will serve to provide insight into some aspects of modelling methodology and will stimulate the reader to contribute to and criticize the literature in this interesting and exciting area of study.

ACKNOWLEDGEMENTS

The authors are indebted to Dr J. Kirk and Dr T. Wheldon (Department of Clinical Physics and Bio-Engineering, West of Scotland Health Board) for introducing them to this absorbing area of study, and for the many subsequent stimulating conversations.

Thanks are due to Elsevier North-Holland Publishers Ltd. for permission to reproduce Fig. 6.3 and to Dr T.T. Puck, Dr P.I. Marcus and the Rockefeller University Press for permission to reproduce Fig. 6.4.

REFERENCES

1. D.A. Abercrombie 'A mathematical model of the growth of cells in culture' (PhD Thesis, University of Strathclyde, 1978)
2. N.M. Bleehan 'Prospects from Radiobiology', *Recent Advances in Cancer and Radiotherapeutics*, ed. K.E. Halnan (Edinburgh and London, Churchill Livingstone, 1972) pp. 217–250
3. G.F. Brunton and T.E. Wheldon 'Characteristic species dependent growth patterns of mammalian neoplasms', *Cell Tissue Kinetics* 11, 161–175 (1978)
4. H. Cramer 'Model building with the aid of stochastic processes', *Technometrics* 6, 133–159 (1964)
5. F. Ellis 'Fractionation in Radiotherapy', *Modern Trends in Radiotherapy*, Vol. 1, eds. T.J. Deeley and C.A.P. Wood (London, Butterworth, 1967) pp. 34–51
6. F. Ellis 'The Relationship of Biological Effect to Dose-time-fractionation Factors in Radiotherapy', *Current Topics in Radiation Research*, Vol. 4, eds. M. Ebert and A. Howard (Amsterdam, North Holland, 1968) pp. 357–397
7. F. Ellis 'Dose, time and fractionation: a clinical hypothesis', *Clinical Radiology* 20, 1–7 (1969)
8. J. Kirk, W.M. Gray and E.R. Watson 'Cumulative radiation effect. Part I. Fractionated treatment regimes', *Clinical Radiology* 22, 145–155 (1971)
9. J. Laurie, J.S. Orr and C.J. Foster 'Repair processes and cell survival,' *British Journal of Radiology* 45, 362–368 (1972)
10. E.L. Powers 'Considerations of survival curves and target theory', *Physics in Medicine and Biology* 7, 3–28 (1962)
11. T.T. Puck and P.I. Marcus 'Action of X-rays on mammalian cells,' *Journal of Experimental Medicine* 103, 653–666 (1956)
12. S.E. Salmon and B.A. Smith 'Immunoglobulin synthesis and total body tumour cell number in I_gG multiple myeloma', *Journal of Clinical Investigation* 49, 1114–1121 (1970)
13. P.W. Sullivan and S.E. Salmon 'Kinetics of tumour growth and regression in I_gG multiple myeloma', *Journal of Clinical Investigation* 51, 1697–1708 (1977)
14. J.R. Usher 'Mathematical derivation of optimal uniform treatment schedules for the fractionated irradiation of human tumours', *Mathematical Bioscience* 49, 157–184 (1980)

15. J.R. Usher and D.A. Abercrombie   'Case studies in Cancer and its Treatment by Radiotherapy' *Int. J. Math. Educ. Sci. Technol.* (to be published)

16. P.F. Verhulst   Loi d'accroissement de la population, *Nuox. Mem. Acad. Roy. Bruxelles* 18(1), 1–38 (1847)

17. T.E. Wheldon and J. Kirk   'Mathematical derivation of optimal treatment schedules for the radiotherapy of human tumours. Fractionated irradiation of exponentially growing tumours', *British Journal of Radiology* 49, 441–449 (1976)

7.

The Humber Tunnel Authority

D.N. Burghes

7.1 STATEMENT OF THE PROBLEM

The Humber Tunnel Authority (HTA) has recently completed a tunnel which links the city of Hull with the expanding North Lincolnshire towns (e.g. Immingham, Grimsby). Because of rising costs, the original plan to build two two-lane tunnels, one for each direction, was changed to a single tunnel with one lane for each direction.

At the moment there is little traffic congestion, but it is expected that in the next decade, work and settlement patterns will change. This will inevitably lead to serious traffic hold-ups at both ends of the tunnel during the morning and evening 'rush hour'.

The Traffic Manager of the HTA has decided to prescribe a recommended speed and separation distance for congested traffic conditions in order to alleviate the expected delays. What recommendations would you advise him to make if he wants to maximise the *flow* of traffic? Should these recommendations apply at all times?

The data overleaf, which might be of help, are taken from the current British *Highway Code*.

Shortest stopping distances

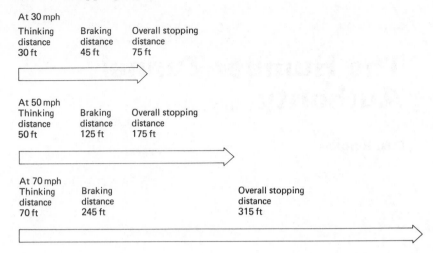

At 30 mph

| Thinking distance 30 ft | Braking distance 45 ft | Overall stopping distance 75 ft |

At 50 mph

| Thinking distance 50 ft | Braking distance 125 ft | Overall stopping distance 175 ft |

At 70 mph

| Thinking distance 70 ft | Braking distance 245 ft | Overall stopping distance 315 ft |

7.2   TEACHING HINTS

This case study has been formulated so that the problem statement (Section 7.1) can be handed out to the class; and it is suggested that, at least initially, no help is given, but that the class is allowed to work on the problem in small groups.

If little progress has been made after one session (e.g. 1 hour, or homework exercise) then it might be a good idea to take the class through to formula (7.1) with suggestions for the separation distance.

The exercise does not require sophisticated knowledge of mathematics. Basic calculus is required, and so it is suggested as a suitable case study for first-year undergraduates, and also as an exercise that could be set early on in a modelling course.

7.3   A POSSIBLE SOLUTION

The important concept to start with is the *flow rate*, which we define as the number of cars passing a fixed point in a unit time interval. The flow rate will depend on a number of factors, such as:

(a)   traffic speed,
(b)   separation distance between cars, and
(c)   length of cars.

For simplicity, we assume a uniform stream of traffic, moving with speed v mph, with an average separation distance d feet between vehicles and an average car length b feet. This is illustrated in Fig. 7.1.

Fig. 7.1   Uniform stream of traffic

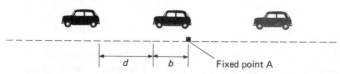

d   b   Fixed point A

We first note that v mph $= 22v/15$ feet per second. Now consider a fixed post, A, and suppose the back of a car has just passed this post (see Fig. 7.1). We want to find the time taken for the next car in the stream to pass A. This is given by

$$\frac{\text{Distance yet to travel}}{\text{Speed}} = \frac{d+b}{(22v/15)}.$$

Now the flow rate f is the number of cars passing A in 1 second;

i.e.   Flow rate $= f = \dfrac{1}{\text{Time taken by each car}}$

$$= \frac{1}{(d+b)/(22v/15)}$$

i.e. $$f = \frac{22v/15}{(d+b)}. \qquad [7.1]$$

Thus the flow rate depends on both speed v and separation distance d, so the traffic manager of HTA has two possible controls: namely, speed v and separation distance d.

However, the speed v clearly affects the distance d. The *Highway Code* data given in Section 7.1 suggest that the shortest stopping distance d_s is made up of two parts:

(a)   the *thinking distance* d_t (which is the distance travelled whilst the driver moves from the accelerator to the brake pedal);

(b)   the *braking distance*, d_b (which is the distance travelled whilst braking from speed v to 0).

From the *Highway Code* data, it is clear that the thinking distance is modelled by

$$d_t = v \text{ feet}. \qquad [7.2]$$

It is not quite so clear what formula has been used for the braking distance d_b, although a graph of d_b against v does show that it is not a linear relationship. If we assume a power law relationship of the form

$$d_b = kv^\alpha \quad (k, \alpha \text{ positive constants})$$ [7.3]

then, taking logs, we obtain

$$\log d_b = \log(kv^\alpha)$$
$$= \log(v^\alpha) + \log k$$

i.e. $\log d_b = \alpha \log v + \log k$ [7.4]

using the properties of logs. If we now consider a graph of $\log d_b$ against $\log v$, Equation [7.4] predicts a straight line graph with slope α and the intercept $\log k$ with the $\log d_b$-axis.

The data from the *Highway Code* are illustrated in a $\log d_b$–$\log v$ graph in Fig. 7.2. We do indeed obtain a straight line, with slope 2. Also, from the intercept with the $\log d_b$ axis, we obtain $k = 0.05$. Hence a possible model used for these data is

$$d_b = \frac{v^2}{20} \text{ feet.}$$ [7.5]

Fig. 7.2   Braking distance–speed relationship

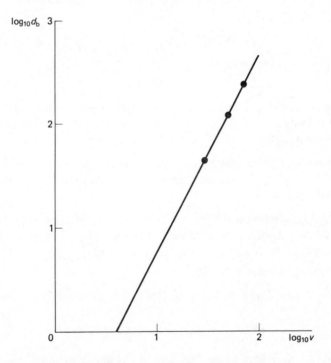

In Appendix 7.1 we give the theoretical reasoning behind the models for d_t and d_b.

We now return to the problem of maximising the flow rate, given in Equation [7.1]. We must decide what form the separation distance, d, takes. The *Highway Code* recommends taking $d = d_t + d_b$, but experience seems to point to more aggressive driving. The other extreme is to take $d = d_t$, just the thinking distance.

We will start by looking at these two extreme cases.

(a) $d = d_t$

In this case

$$f = \frac{22v}{15(v+b)} \tag{7.6}$$

and this is illustrated in Fig. 7.3.

Fig. 7.3 Flow rate for $d = d_t$ $(b = 12 \text{ feet})$

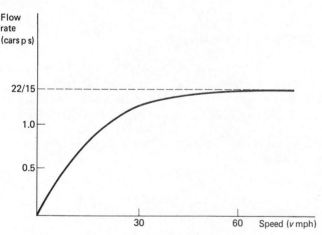

We see that $f \to 22/15$ as $v \to \infty$, and the policy seems to be 'the faster the better'! This is certainly not the policy to be encouraged by the HTA. It is interesting to note though that the flow rate only increases very slowly above 30 mph.

(b) $d = d_t + d_b$

Here the flow rate is given by

$$f = \frac{22v}{15(v + v^2/20 + b)}. \tag{7.7}$$

The behaviour of this function is illustrated in Fig. 7.4.

Fig. 7.4 Flow rate for $d = d_t + d_b$ $(b = 12 \text{ feet})$

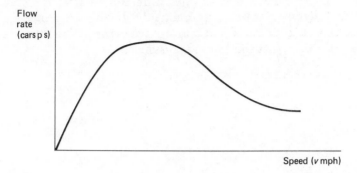

This function has a maximum value at a finite value of v, given by the solution of

$$\frac{df}{dv} = 0$$

i.e. $(v + v^2/20 + b) = v(1 + v/10)$.

Solving gives $v = \sqrt{20b}$ and substituting $b = 12 \text{ feet}$, we obtain $v \approx 16 \text{ mph}$. The corresponding separating distance being 28 feet. Again, this extreme case is probably not a very realistic one. Most drivers will want to drive faster than 16 mph, and closer than 28 feet!

To look for a situation somewhere between these two extremes, we take the separation distance as

$$d = d_t + \alpha d_b$$

where α is a parameter in the range $0 < \alpha < 1$. In this case

$$f = \frac{22v}{15(v + \alpha v^2/20 + b)} \qquad\qquad [7.8]$$

and putting $df/dv = 0$ gives

$$v + \alpha v^2/20 + b = v(1 + \alpha v/10)$$

i.e. $v = (20b/\alpha)^{1/2}$,

the corresponding separation distance being

$$v + \alpha v^2/20 = v + b = (20b/\alpha)^{1/2} + b.$$

For $b = 12 \text{ feet}$, Table 7.1 illustrates how the factor α affects the optimal speed and separation distance.

Table 7.1 Dependence of optimal speed and separation distance on parameter α

α	Optimal speed v (mph)	Corresponding separation distance d (feet)
1.0	15	27
0.75	18	30
0.67	19	31
0.5	22	34
0.25	31	43

A possible solution for the Traffic Manager of HTA is to display a notice board at the entrance of the tunnel stating

> IN CONGESTED TRAFFIC
>
> TRAVEL AT 20 mph
>
> SEPARATION DISTANCE 11 YARDS

One important point to note is that as well as specifying the optimal speed, the recommended separation distance must also be given, otherwise operating conditions may be far from optimal!

7.4 RELATED PROBLEMS

1. How sensitive are the results obtained in Section 7.3 to:
   (a) the value used for b, the average car length;
   (b) the value used for the thinking time (see Appendix).

2. What suggestions would you make (if any) for regulating the flow of congested traffic on a multilane motorway?

3. Study a street map plan of your nearest large town or city. Can you suggest a different system of road traffic flow (but using the existing network) which would improve the efficiency of the network?

APPENDIX 7.1

(a) Thinking Distance

If t is the thinking time in seconds, then the thinking distance, assuming constant speed v m.p.h., is

$$d_t = \left(tv \frac{22}{15} \right) \text{ft}$$

In order to arrive at a simple formula, the Ministry of Transport uses $t = 15/22$ seconds to give Equation [7.2],

$$d_t = v \text{ feet.}$$

The thinking time will of course vary from person to person; a grand-prix driver will have a value less than $15/22$, whilst elderly drivers could have values well in excess of $15/22$.

The important point is to stress that as the speed increases so does the distance travelled, and this simple formula illustrates this.

(b) Braking Distance

As above, the emphasis here is again on simplicity rather than exactness. We start by assuming that the maximum braking force is $\frac{2}{3} mg$ (m being the car's mass). This is of course just an average value. Now, if x denotes distance travelled, then

$$\frac{d^2x}{dt^2} = -\frac{2}{3}g.$$

Integrating gives

$$\frac{dx}{dt} = -\frac{2}{3}gt + v$$

which on integrating again gives

$$x = -\frac{1}{3}gt^2 + vt.$$

We have $x = d_b$ when $dx/dt = 0$ which gives $t = 3v/2g$. Thus

$$d_b = \frac{3v^2}{4g}.$$

To obtain d_b in feet, we must write the speed as $22v/15$ (with v in mph) and g as $32.2 \, \text{ft s}^{-2}$. This gives

$$d_b = \frac{3(v22/15)^2}{4(32.2)} = \frac{v^2}{19.96} \text{ ft}$$

which is approximated by

$$d_b = \frac{v^2}{20} \text{ ft.}$$

The important factor here is the v^2 dependence rather than the exact numerical value. This is why the modelling has not been precise.

RELATED READING

1. D.N. Burghes   'Mathematical Modelling: A Positive Direction for the Teaching of Applications of Mathematics', *Educ. Studies Math.* (1980)
2. R. Haberman   *Mathematical Models* (New Jersey, Prentice-Hall, 1977)
3. Open University, *Course TM 281*, 'Modelling by Mathematics' Units (Milton Keynes, Open University, 1978)

8.

Parking a Car

S.C. Dunn

8.1 INTRODUCTION

Private motor transport is now very widely used. The car represents a considerable personal investment, and its continued availability is an important concern to the owner. When a car is driven on the road, the instinct of self-preservation on the part of the driver will tend to keep that car from obvious harm. However, when manoeuvring in confined spaces and particularly when parking, the driver must often show great skill if the remarkably fragile fabric of the car is not to be reshaped rather arbitrarily. Such skill depends on sensor–motor responses which differ from driver to driver, but the ultimate limit to permissible movement depends on the mechanical construction of the car itself.

The steering mechanism of a car causes the two front wheels to pivot about vertical axes through each wheel. These vertical axes pass through the horizontal axis of rotation of each wheel and the effective contact point of the wheel with the road. We say 'effective' contact point because the tyre is locally flat where it meets the road, and when the car is steered the contact area rotates about a point within the area. What is more important is that the two wheels do not pivot by the same amount. They move so that their horizontal axes always intersect on the line through the rear axle. Fig. 8.1 illustrates this arrangement.

As the steering wheel is turned, the point X, the rear-axle interception, travels along the axle line. The limit of steering capability is reached when the front wheels have pivoted their utmost. The distance between X and the further rear wheel road contact 'point' may be quoted as the minimum turning radius for the car. A knowledge of this value of radius would enable us, for example, to determine the least width of street in

Fig. 8.1   Steering mechanism of a car

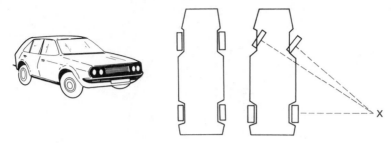

which it would be possible to do a U-turn without mounting the pavement. A problem which is less directly solvable is the set of manoeuvres needed to park and 'un-park' a car when surrounded by other vehicles.

At this stage I think the students should be asked to consider on their own how they would remove the subjective element from the manoeuvring process and reduce it to a purely objective analysis. This will involve the identification of the simplifications necessary to help the analysis.

Other topics at this point could include:

(a)   the diverse factors of interest when moving a car among obstacles;

(b)   an assessment of the need to re-introduce later the observations made by a driver when actually at the steering wheel;

(c)   identification of the measurements which would have practical value.

When general discussion is resumed there will probably be a consensus of opinion about what constitutes good driving practice but less certainty about how to proceed with analysis and for what purpose. Experienced drivers know that a space in a line of parked cars can only be neatly entered and occupied by 'backing in'. Driving in forwards requires much more room. Why this should be so can be best appreciated practically by unparking a neatly stowed car, and graphically with the aid of a simple template. (By unparking we mean driving the car out of the space in the opposite sense to which it was driven in.)

A suitable template is shown in Fig. 8.2. It can be made, for example, from the clear polythene closure supplied for use with a tin of ground coffee. The aperture is of such a size that a sharp pencil will trace out on paper a rectangle whose corners correspond to the points of road-contact of the four wheels. To the same scale as that trace, two small holes are positioned representing the centres about which the car turns at minimum radius in either direction. A pin pushed through one of the holes forms a spindle about which the template can be turned on the surface

Fig. 8.2 A suitable template

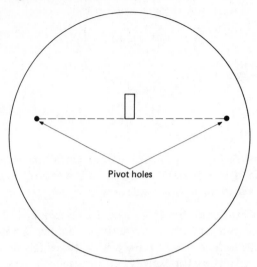

Pivot holes

of a drawing sheet and a series of figures can be drawn showing the successive positions taken up by a manoeuvring car.

The four parts of Fig. 8.3 show the ideal movements in parking a car by backing into a space formed by a gap in a line of cars. Needless to say, the author cheated by drawing the diagrams in the reverse sequence!

This stage in the reduction of the problem corresponds, I suggest, to the exploration of algorithms in computer program writing. I believe it is advisable to spend quite a long time getting benefit out of the graphical method of studying different situations before 'encoding' these situations in analytical form. For example, at this level of appreciation other manoeuvres should be examined. Fig. 8.4 shows the disadvantage of the 'drive-in forwards' approach. Honesty has prevailed here and the vehicle is progressively unparked, riding up over the kerb-line as it withdraws.

Fig. 8.3 Parking a car by backing into a space

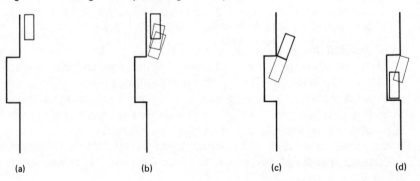

(a)   (b)   (c)   (d)

The advantage of simply 'playing' with the graphical aid is that it will often point to a more significant restatement of a problem, offering a more useful answer.

Fig. 8.4   Drive-in approach to parking

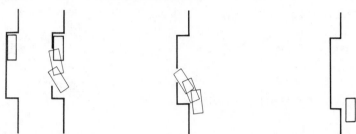

A problem which drivers often face is the need to extricate a car from a space which, since the time of parking, has changed its dimensions, due to the movement of other cars. It may also be required at times to move a car nearer the kerb without quitting the space.

If circumstances permit it is a good idea to have a further period of relatively unsupervised study with the stated objective of recommending the kind of information a closer examination of the problems should provide. The aim should be to reach general agreement that it would be useful to derive generalised values for parameters such as those in Fig. 8.5.

Fig. 8.5   Parameters involved

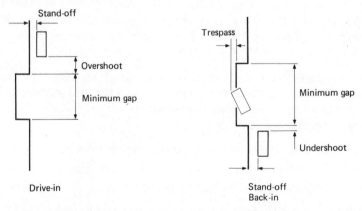

Use of the template should have made students familiar with the idea of approximating the behaviour of the real car by neglecting the dimensions of the wheels and the superincumbent bodywork. Nevertheless it is good practice to relate the simplified model to a real object. Fig. 8.6 shows the essential dimensions of a British Leyland *Princess 2* together with the representation made of it.

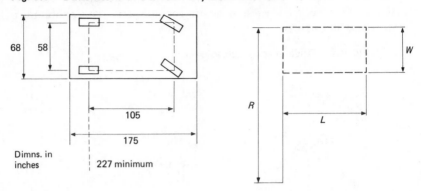

Fig. 8.6 Dimensions of a British Leyland *Princess 2*

68 | 58

105

175

Dimns. in inches

227 minimum

R

L

W

The discipline should be inculcated of writing down in words or compiling tables which describe exactly the sequence of positions and movements which it is intended to put in symbolic form. For the sequence of Fig. 8.3 the following would be appropriate.

8.2 BACKING IN

In Fig. 8.3(a) a car has driven past a suitable space at the left-hand kerb; in Fig. 8.3(b) the steering wheel has been turned fully to the left and the car reversed till the nearside rear corner is adjacent to the car parked in front. The steering wheel is returned to 'dead centre' and the car backed by its own length as in Fig. 8.3(c). The steering wheel is turned fully in the opposite direction and reversing continues. If the initial placing of the car was judged correctly, it will now be possible to stop the car parallel to the kerb and just touching it as in Fig. 8.3(d).

Given the dimensions of the car represented by Fig. 8.6 (length L, width W, and minimum turning radius R) it is a straightforward matter to calculate three other dimensions which are of interest to the driver and shown in Fig. 8.5, namely:

(a)   the minimum length of space in which it is possible to park;

(b)   the amount by which one should 'overshoot' the space; and

(c)   the 'stand off' from the line of parked cars.

It will be assumed that all the other cars have been parked perfectly and are of the same width; hence the width of the parking space is W. It is convenient to follow the motion of the car by unparking it, assuming it just fails to touch the car behind it.

The car is driven forward at 'full-lock' till the near-side front corner is at a distance W from the kerb (and just misses the car in front). The geometrical relations are set out in Fig. 8.7. The effective turning radius of the near-side front corner is $\sqrt{R^2+L^2}$ and the angle between that radius and the original position of the rear axle is

$$\beta = \frac{\cos^{-1}(R-W)}{\sqrt{R^2+L^2}}$$

The distance ab measures the minimum parking length (since the car is driven straight out in the next phase) and is $(R-W)\tan\beta$.

Fig. 8.7 Geometrical relations when unparking

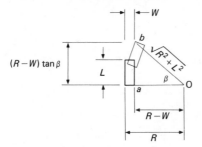

With the steering wheel centred, the car is driven forward by its own length and on full, opposite lock is taken to the parallel, offset position in Fig. 8.8. The angle through which the car rotates in this last manoeuvre is $\beta-\alpha$ where α is the angle subtended by the near-side car length at the pivot point. By symmetry this is the same angle through which the car swings to reach the drive-out position. The overshoot is $(R-W)\sin(\beta-\alpha)$.

Fig. 8.8 Rotation to drive-out position

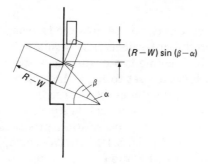

The stand-off can be found from a similar examination of the triangle ebc which, for clarity, has been reproduced in Fig. 8.9. The segment

$$ed = (R-W)\cos(\beta-\alpha)$$

and the segment

$$ec = R-W$$

so that the stand-off

$$dc = ec - ed = (R - W)(1 - \cos\overline{\beta - \alpha}).$$

For the model car of Fig. 8.6 ($L = 105$, $W = 58$, $R = 227$), the minimum gap in which one may park is longer than the car by three quarters of its length. To park, first overshoot by 62% of car length and stand-off by about a quarter of its width.

Whether the foregoing example is worked out as a demonstration or set as an exercise is a matter of choice. The next should certainly be treated as student work.

Fig. 8.9 Drive-out position

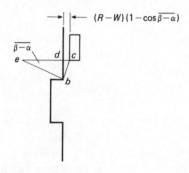

8.3 DRIVING IN

There are circumstances in which it may not be possible to overshoot the space and the car must be entered front-wheels first. Again with the ruse of unparking the car, this time by backing it out, the successive positions taken up are in Fig. 8.10. The wheels are set to full right lock and reverse gear is engaged. As it moves backwards the kerbline is crossed. When the nearside corner of the car just reaches one car-width from the kerb (Fig. 8.10(a)) full left lock is substituted and motion proceeds until a parallel, unparked position is occupied (Fig. 8.10(b)). The angle, through which the car rotates first anti-clockwise then in the opposite sense is $\cos^{-1}(1 - W/R)$. The amount by which the minimum gap must exceed the length of the car is $R \sin\gamma$. Where an algebraic expression is preferred

$$R \sin\gamma \equiv \sqrt{2RW - W^2}.$$

From Figs. 8.10(b) and (c) it will be seen that a driver who is able to cross the kerb line may park exactly if he stops short of the gap by $R \sin\gamma - L$ and stands-off by $(R - W)(1 - \cos\gamma)$ or $W(R - W)/R$.

Fig. 8.10 Driving-in position

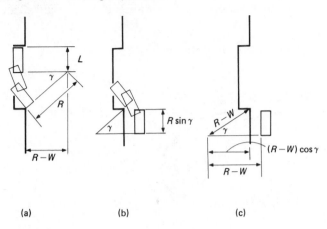

(a)   (b)   (c)

For our model this means a minimum parking space a little less than $2\frac{1}{2}$ times the length of the car, stopping short of the gap by almost half a length and standing out by about three quarters of the width. However, the kerb line is trespassed by $R(\sec\alpha - 1)$, corresponding here to about 40% of the car's width.

By this stage, one hopes that some of the students will have derived analytical expressions and most will have mastered the graphical treatment. Irrespective of the level of achievement there should at least be opinions forthcoming about the formal result of 'driving in' and its relationship to real driving practice.

Not only is kerb-crossing undesirable, in many cases, where parking is against a wall for example, it is not even possible. Is parking by driving forwards under these circumstances effective?

To answer this question, as before, we require a description in words or by a clear diagram of exactly what the experienced driver does when he has no alternative but to make the forward approach. It is most important to spend enough time formulating this description; if it is not sufficiently comprehensive the crucial step of translating it into a relationship between variables may not be taken at all. A typical description might be the following.

If a car is to be parked against a pavement by driving in frontwards, the optimum approach begins with driving in at such an angle that when the wheels are about to touch the pavement they can be 'put on maximum lock' and are then just parallel to the pavement. Thereafter as the car moves forwards, the steering wheel is wound back from its limiting position so as to maintain the wheel next to the pavement always parallel to it (i.e. its axis is always perpendicular to the line of the pavement).

In the next stage the operation is reduced to two wheels freely rotating on axles appropriately spaced and connected to a rigid frame, as in Fig. 8.11.

Fig. 8.11 Two freely rotating wheels

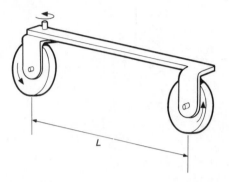

It may prove necessary to make a physical model of this linkage in order to register its basic property, that of rotation about a fixed centre. It may even be worthwhile making the rear wheel in two parts side by side in a common trunnion so that a ball-point pen can be accommodated between them to actually draw a locus.

The next stage is crucial. We ask what further simplification is possible? It is not too fanciful to say that hardware is replaced by software. The wheels are removed and their functions replaced by descriptions and the kerb is represented by a straight line. The rigid frame is now a line-segment, and hence it is easy to substitute for the action of the front wheel against the kerb: 'one end of the line-segment moves along the kerb-line'. What is not so obvious is that the other end of the line-segment has a constrained motion. In the physical model this happens because there is no force causing sideways motion and the rear wheel can only roll in the direction pointed by the rigid frame. In the word-model we can summarise the motion as follows.

In Fig. 8.12 let the y-axis represent the kerb-line, the line-segment L the line joining the centres of the front and rear wheels on the side of the car nearer the kerb. The angle which L makes with y-axis we will call θ. L moves in such a way that one end travels along the y-axis while the other end moves in the direction of L.

Before proceeding further the student should be asked to comment on the respective ranges of operation over which the model and original are valid. Because the 'front wheel' of the model can be freely pivoted without limit, the line-segment L can successively and continuously take up positions, a few of which are shown in Fig. 8.13. What are the bounds to the original motion?

Fig. 8.12 Simplified model

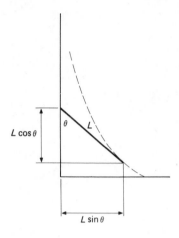

$L \cos \theta$

θ L

$L \sin \theta$

Fig. 8.13 Possible positions for line segment L

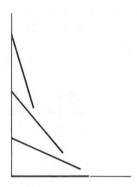

The trailing end of L moves along a defined path and the student can be asked for the equation of that path. It will usually be found that attempts are made simply to derive from the geometry an expression for y in terms of x. These bare trials should not be discouraged, but eventually hints may be given about making some use of θ. The extrication of a useful result is in two parts.

In the first we reduce the problem to an expression for the slope of the curve traced out by L. At a point on the curve whose abscissa is x the slope is that of the line L. Therefore

$$\frac{\mathrm{d}y}{\mathrm{d}x} = \cot \theta.$$

A memory of trigonometric relations should prompt a search for a cosecant

$$L/x = \operatorname{cosec} \theta$$

and since

$$\cot^2\theta = \operatorname{cosec}^2\theta - 1$$

$$\left(\frac{dy}{dx}\right)^2 = \frac{L^2}{x^2} - 1$$

whence

$$dy = \pm \sqrt{\frac{L^2}{x^2} - 1}\, dx$$

and

$$y = \pm \int \left(\frac{L^2}{x^2} - 1\right)\, dx \qquad\qquad [8.1]$$

the constant of integration being zero by inspection.

In the second part a standard table of integrals[3] is consulted. There are two routes to the answer. Firstly, direct use of Equation [8.1] leads to

$$\pm \frac{y}{L} = \sqrt{1 - \frac{x^2}{L^2}} + \ln\left|\frac{1 - \sqrt{1 - (x^2/L^2)}}{x/L}\right| \qquad\qquad [8.2]$$

which may also be expressed in terms of inverse hyperbolic functions. Secondly, on reverting to the variable θ, Equation [8.1] becomes

$$\frac{y}{L} = \pm \int \cot\theta \cos\theta\, d\theta$$

and referring to our table of integrals gives

$$\pm \frac{y}{L} = \cos\theta + \ln\left|\tan\frac{\theta}{2}\right|. \qquad\qquad [8.3]$$

The latter form resolves the sign ambiguity neatly, for $\tan \theta/2$ will always be positive and less than unity, hence $\ln|\tan \theta/2|$ will thus always be negative and, for practical values of θ, numerically greater than $\cos\theta$. Hence the negative sign on the left-hand side of Equations [8.2] and [8.3] prevails.

The curve is asymptotic to the kerb line and is shown in Fig. 8.14. (This curve is of course that known in old text books as the tractrix. It also arises in the analysis of the wear of footstep bearings.[1] A formula is quoted in Reference 2.)

Fig. 8.14 Path of trailing end

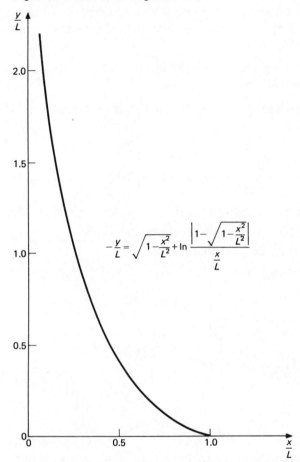

$$-\frac{y}{L} = \sqrt{1-\frac{x^2}{L^2}} + \ln \frac{\left|1-\sqrt{1-\frac{x^2}{L^2}}\right|}{\frac{x}{L}}$$

8.4 VALIDATION

Compared with some modelling problems validation of the methods described here are almost trivial because the analysis is, in principle, no more than a process inverse to that used to design the motor car originally. Although practical work is not to be discourged, it does not take much more than the inspection of a real car to reveal that the only refinement which might be advisable is a correction for the extra space taken up by the body.

8.5 FURTHER WORK

What has been provided so far is a graphical method to be used when turns are made at a fixed radius and a formula usable when the steering is varied according to a fixed law. The study of circumstances more complicated than those so far presented encourages another valuable skill, the organizing of a composite solution by synthesis of a number of separate, partial answers.

Other situations of practical interest include the following. The first group are limited to 'driving in'.

1. The safest possible entry begins when your rear wheels are adjacent to the front wheels of the car you will park beyond. Full steering lock first in one direction then the other leaves the car parallel to the kerb but stood out some way (Fig. 8.15).
2. The same entry but after touching the kerb the road wheel is kept parallel to it (Fig. 8.16).
3. As 2. but finishing parallel by a short final full lock turn (Fig. 8.17).
4. As 1. but reversing to restore neatness (Fig. 8.18).
5. As 2. but finished like 4. (Fig. 8.19).
6. As 3. but finished like 4. (Fig. 8.20).

Fig. 8.15–8.20 Various driving-in manoeuvres

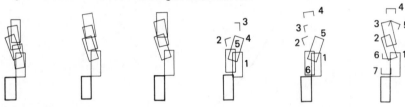

The second group are awkward manoeuvres.

7. The classic extrication manoeuvre needed when you return to find yourself 'boxed in' (Fig. 8.21).
8. Escape from a cul-de-sac where the approach to the end-space is almost denied by other cars (Fig. 8.22).
9. The herring-bone torture (Fig. 8.23).
10. The public car-park (Fig. 8.24).

Fig. 8.21 Manoeuvres when 'boxed in'

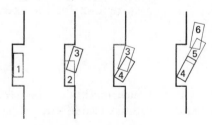

Fig. 8.22   Escape from a cul-de-sac

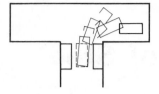

Fig. 8.23   The herring-bone torture

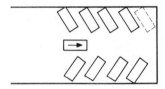

Fig. 8.24   The public car park

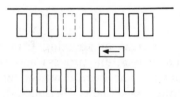

A question which will no doubt arise is 'how applicable is the graphical analysis to the real situation where the driver must choose his movements on the basis of his much poorer view of the relative positions of car and obstacle?' Observation seems to support the view that the driver assumes that the effective space for manoeuvre is smaller than what is actually available so that a margin is left for error of judgment as well as body size.

REFERENCES

1. W.N. Rose   *Mathematics for Engineers*, Part II (London, Chapman and Hall, 1945) p. 333
2. H.J. Bartsch   *Mathematische Formeln* (Koln, Buch-und Zeit-Verlagsgesellschaft mbH, 1975) p. 308
3. H.B. Dwight   *Tables of Integrals and Other Mathematical Data* (New York, Macmillan, 1947) p. 63

9.

Air Gap Coiling of Steel Strip

I.D. Huntley

9.1 INTRODUCTORY BACKGROUND

The following problem arose when a light engineering firm became dissatisfied with wastage rates in its production process and required some assistance in analysing the problem. The firm, however, wishes to remain anonymous, and so certain details have had to be omitted.

The problem can be tackled on many different levels, using as little or as much mathematics as required. It can certainly be tackled by pre-calculus students.

The firm in question manufactures plain bearings — steel cylindrical shells coated with a bearing material — whose purpose is to reduce sliding friction on engine components (see Fig. 9.1).

The steel is purchased as a coiled strip which is then uncoiled and coated with the bearing material before being recoiled and placed in an oven for curing. In order that the curing is effective it is essential that an air gap is maintained between adjacent coils, and so during the recoiling operation chains are fed onto both edges of the strip to act as spacers. The chains are held in place by the coiling tension and the securing bands (see Figs. 9.2 and 9.3).

Unfortunately, problems have been encountered in the production of the continuous air gap between the coils using this method:

(a) considerable safety precautions are necessary at the recoiling station;

Fig. 9.1 Plain bearings

Fig. 9.2 Chain dispensing and coiling

Fig. 9.3 Steel strip coiled using the chain as a spacer

(b)   chain interleaving is labour-intensive;

(c)   the chain is expensive to replace when breakages occur;

(d)   the chain tends to wander from the selvedge (the waste portion at the edge which is later cut off) onto the central part of the strip (see Fig. 9.4), thus reducing the usable bearing material width;

(e)   the chain tends to fall out, causing collapse of the air gap (see Fig. 9.5).

Thus the firm has been investigating several other methods for producing the continuous air gap.

After considering various alternatives, the firm decided to develop a method which uses a rotary tool to impart intermittent nicks, which act as spacers, to both edges of the steel strip (see Figs. 9.6 and 9.7).

Fig. 9.4   Scrap strip created when the chain wanders on to the bearing material

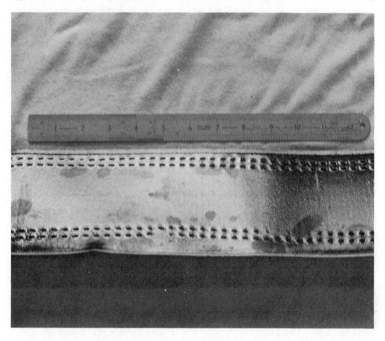

Fig. 9.5   Scrap strip caused by collapse of the air gap

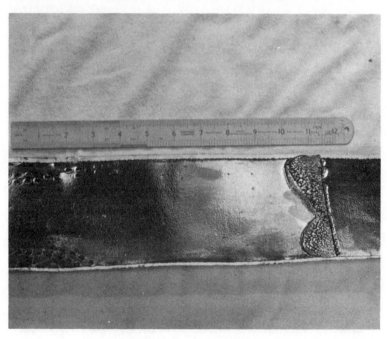

Fig. 9.6 Rotary tool and strip with nicks

Fig. 9.7 Nick pitch and nick coincidence

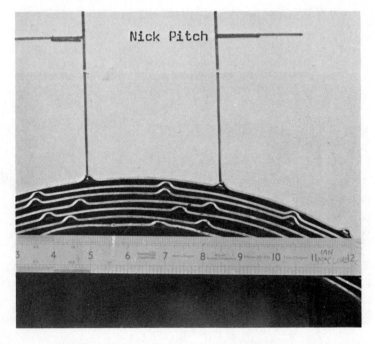

9.2   STATEMENT OF THE PROBLEM

If the air gap between the coils of the steel strip collapses during the curing process, a great deal of wastage results. Preliminary experiments have shown the firm that nick pitch, the distance round a coil between adjacent nicks, is an important variable:

(a)   a large nick pitch means a large unsupported length of coil and air gap collapse through the coil sagging into a polygon shape (see Fig. 9.8);

Fig. 9.8   Polygon-shaped coiled strips

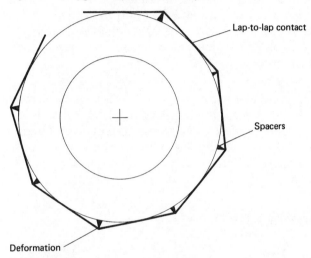

Lap-to-lap contact

Spacers

Deformation

(b)   a small nick pitch leads to higher stress concentrations in the strip and an increased probability of gap collapse because of nicks on adjacent coils coinciding (see Fig. 9.7).

In fact, for the actual dimensions used, experiments have shown that a nick pitch greater than 15 cm results in the polygon shape and consequent air gap collapse, and that each collapse of the air gap results in about 15 cm of waste strip where the bearing material is damaged. Typical dimensions are:

Strip thickness	2 mm
Strip width	100 mm
Strip length	50 m
Diameter of coiling former (i.e. diameter of innermost coil)	380 mm
Air gap size	6 mm

The problem, as posed by the firm, is theoretically to determine the optimum nick pitch, since a comprehensive range of experiments would be too costly. We proceed on the assumption that the pitch is less than the 15 cm limit quoted above, and that the wastage is directly proportional to the number of nick coincidences.

9.3 MODEL 1

Before rushing into mathematical formulations of the problem, it is often useful to see if sketches and diagrams help at all. It is clearly not difficult to make a scale drawing of a coiled strip and use this to investigate the number of nick coincidences (or near coincidences) for various different nick pitches.

An example of this is shown in Fig. 9.9, where it is clear that if the strip remains flat between the supporting nicks — that is, follows the tangent line marked — a nick pitch greater than about 15 cm will result in the collapse of the air gap prophesied by the experiments. The method, however, is extremely tedious.

Fig. 9.9 Scale drawing of part of coiled strip

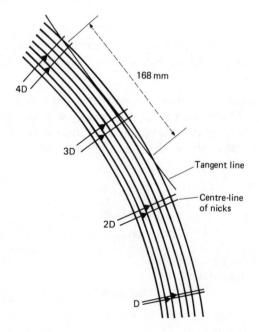

9.4 MODEL 2

Since even the innermost coil of the strip is of considerable diameter, it would seem not too serious an approximation to assume that each coil is in fact a circle. The radius of each lap increases by the 6 mm air gap plus the material thickness, and so we can calculate the (constant) lap length difference d. In the real coiled strip most of the variables are slowly changing, so to make sense we must now consider averaged values of all the variables. The average number of nicks per lap $\overline{N}$ is obtained by dividing the average lap length by the nick pitch P:

$$\overline{N} = \frac{2\pi(r_{min} + r_{max})}{2P}$$

where r_{min} is the radius of the inner coil and r_{max} that of the outer coil (which can be found by trial and error based on circular coils). It is clear from the scale drawing in Fig. 9.9, that each nick moves a constantly increasing distance D relative to the corresponding nick on an adjacent lap. The value of D decreases as the coil diameter increases, so we may define an average value $\overline{D}$ by dividing the lap length difference by the average number of nicks per lap:

$$\overline{D} = \frac{d}{\overline{N}}$$

Then the average number of nicks between two coincidences is obtained by dividing the nick pitch P by the constant $\overline{D}$, and the average length $\overline{L}$ between coincidences is this number multiplied by P:

$$\overline{L} = \frac{P^2}{\overline{D}}.$$

Finally, the average number of coincidences per coiled strip is coil length/$\overline{L}$ and the total scrap length is easily obtained.

Experience has shown that students readily approximate the coiled strip by concentric circles but have difficulty thereafter. Thus a discussion with the teacher is often valuable at this stage.

Table 9.1 shows the way these data may be presented, from which one would clearly choose to use the largest pitch and expect something around 3% scrap per coiled strip. Alas, for a similar geometry, the preliminary experiences had indicated about 0.1% scrap! An obvious source of error is that it may not be necessary for the nicks exactly to coincide in order to get air gap collapse — but allowing for this fact would increase the scrap and not decrease it! Thus we now try a geometry that more closely resembles the coiled strip.

Table 9.1 Predicted scrap levels due to nick coincidences

Thickness (mm)	Pitch P (mm)	$\overline{L}$ (m)	Number of coincidences	Scrap (m)	Scrap (%)
2	50.8	1.88	26.6	3.99	8.0
	76.2	2.82	17.7	2.66	5.3
	101.6	3.76	13.3	2.00	4.0
	127.0	4.70	10.6	1.60	3.2
	152.4	5.64	8.9	1.33	2.7
3	50.8	1.78	28.1	4.22	8.4
	76.2	2.67	18.7	2.81	5.6
	101.6	3.56	14.1	2.11	4.2
	127.0	4.45	11.2	1.69	3.4
	152.4	5.33	9.4	1.41	2.8

To the accuracy shown here, the 'circle method' of Model 2 and the 'spiral method' of Model 3 gives identical results.

9.5 MODEL 3

Since the coiled strip is of spiral form, we now consider equations for a spiral. The two most obvious spirals are the equiangular spiral $r = \alpha^{\theta}$ and the Archimedean spiral $r = \alpha\theta$, where r is the radius and θ the angle. To see which of these best fits the coiled strip, we consider the radii for angles $\theta = 0, 2\pi, 4\pi, \ldots$. These are given in Table 9.2.

Table 9.2 Radii for various values of θ

Angle	Equiangular spiral	Archimedean spiral
0	1	0
2π	$\alpha^{2\pi}$	$2\pi\alpha$
4π	$\alpha^{4\pi} = (\alpha^{2\pi})^2$	$4\pi\alpha = 2(2\pi\alpha)$
6π	$\alpha^{6\pi} = (\alpha^{2\pi})^3$	$6\pi\alpha = 3(2\pi\alpha)$
.	.	.
.	.	.
.	.	.

Clearly only the Archimedean spiral can represent a situation where the distance between adjacent laps of the coiled strip is constant, and so we consider $r = \alpha\theta$.

Elementary calculus texts tell us that the arc length between θ_1 and θ_2 is given by

$$S = \int_{\theta_1}^{\theta_2} \sqrt{r^2 + \left(\frac{dr}{d\theta}\right)^2}\, d\theta$$

$$= \alpha \int_{\theta_1}^{\theta_2} \sqrt{1 + \theta^2}\, d\theta, \quad \text{in the case} \quad r = \alpha\theta$$

i.e. $\quad S = \dfrac{\alpha}{2} \{\theta\sqrt{1 + \theta^2} + \ln(\theta + \sqrt{1 + \theta^2})\} \Big|_{\theta_1}^{\theta_2}$ [9.1]

from standard tables of integrals.

We now use formula [9.1] to repeat the calculations of Table 9.1:

Lap length difference $= d$

Average number of nicks per lap $= \overline{N} = \dfrac{S_{first} + S_{last}}{2P}$

where S_n is the circumference of the nth lap,

Average increasing distance $= \overline{D} = \dfrac{d}{\overline{N}}$

Average length between coincidences $= \overline{L} = \dfrac{P^2}{\overline{D}}$

Average number of coincidences per coiled strip $= \dfrac{\text{coil length}}{\overline{L}}$

This calculation is also demonstrated in Table 9.1, and it is at once clear that the 'circle method' and 'spiral method' give identical results. Thus we have demonstrated that — when considering averages — it is all right to treat the coiled strip as a set of concentric circles, but we have not thrown any light on the slight discrepancy between theory and experiment.

9.6 MODEL 4

A major advantage for the spiral method is that — in conjunction with a small computer — it gives us a method for solving the problem exactly. Taking Equation [9.1] above, we may write

$$\frac{2P}{\alpha} = (\theta\sqrt{1 + \theta^2} + \ln(\theta + \sqrt{1 + \theta^2})) \Big|_{\theta_n}^{\theta_{n+1}}$$

where θ_n is the (known) position of one nick and θ_{n+1} is the (unknown) position of the next nick. So, by a simple numerical analysis, we may obtain θ_{n+1} to the accuracy required for any choice of pitch P. Repeating the procedure, we build up a list of angles θ_n at which the nicks occur. To examine nick coincidences, we then transform the angles so that $0 \leqslant \theta_n < 2\pi$ and look for duplication on adjacent laps.

This task is made significantly easier if it is noted that

$$\theta\sqrt{1+\theta^2} + \ln(\theta + \sqrt{1+\theta^2}) \doteq \theta^2$$

for typical values of θ. Thus, having worked out α and θ_1, we choose the required pitch P and use the iterative formula

$$\theta_{n+1}{}^2 = \frac{2P}{\alpha} + \theta_n{}^2$$

or $\quad \theta_{n+1} = \theta_n + \dfrac{P}{\alpha\theta_n}$

to generate the required list of angles.

The alert student may note here that the latter is exactly the formula generated by assuming that the coil is circular, but this in itself is a valuable piece of information about Model 2.

Table 9.3 gives the values generated in this way, where the last column shows θ_n in the range $[0, 2\pi]$. It is then a simple matter to pick out those angles where a nick coincidence occurs. The program is then rerun for different values of the pitch P and the pitch leading to the least coincidences is chosen as optimal.

The method is also easily extended to the case of 'near coincidence' where two nicks do not quite coincide but still lead to air gap collapse. The 'near coincidence' may be defined, for any lap, via an angle ϵ, and the list θ_n searched for $\ |\theta_i - \theta_j| < \epsilon \ $ on adjacent laps.

9.7 FURTHER QUESTIONS

1. The coiled strip can be loaded into the curing oven with its axis either horizontal or vertical. Which is best?

2. We have not mentioned the width of the nicks themselves. What factors influence this?

3. What effect would a larger radius former have?

4. We mention above that the nicks are produced by the rotary tool shown in Fig. 9.6. How does the speed of this tool effect the total production process? Will the nick pitch be constant in practice?

TABLE 9.3 Numerical Output

APPROXIMATE METHOD FOR SPIRAL

--- --- --- --- --- --- --- --- ---

INPUT NUMBER OF VALUES REQUIRED
30

FOR NICK PITCH = 50.8 MM AND
STARTING FROM THETA = 149.25, VALUES ARE

1	149.52	5.00
2	149.78	5.27
3	150.05	5.54
4	150.32	5.80
5	150.58	6.07
6	150.85	0.05
7	151.11	0.32
8	151.38	0.58
9	151.64	0.84
10	151.90	1.11
11	152.17	1.37
12	152.43	1.83
13	152.69	1.89
14	152.95	2.15
15	153.21	2.41
16	153.47	2.68
17	152.73	2.94
18	153.99	3.19
19	154.25	3.45
20	154.51	3.71
21	154.77	3.97
22	155.03	4.23
23	155.29	4.49
24	155.54	4.74
25	155.80	5.00
26	156.05	5.26
27	156.31	5.51
28	156.56	5.77
29	156.82	6.02
30	187.87	6.28

5.   How would the above models be changed if certain design tolerances were included. For instance, if the steel thickness were 1.78 mm ± 10% how would this effect the models?

6.   The above scenario supposes that the firm has already decided to use a particular shape of nick. What effect does nick shape have? Can you think of a nick shape for which nick overlap would present no problem.

9.8   COMMENT

As presented above, as a question and four answers, this would probably not be the best way to proceed in the classroom or lecture-room. The manner of presentation would clearly depend on the ability of the class, but a suggested plan is the following.

1. Distribute general background and relevant Figures.

2. Leave (for $\frac{1}{2}$ hour?) for students to think about the problem.

3. Discuss the problem in general terms, ensuring that everyone has a clear understanding of the situation and throwing out a few hints like 'pictures', 'circles', 'spirals'.

4. Give the class a week, and ask for a written answer — preferably a non-technical report and a technical appendix.

5. Ask two or three students to present their models to the rest of the class.

10.

Hydroelectric Power Generation System

D.J.G. James

10.1 LEVEL OF DIFFICULTY

For a detailed study a thorough knowledge of constant coefficient second-order linear differential equations is required. Consequently, the case study is appropriate for second-year undergraduates. However, the study of a simplified version requires only a knowledge of first-order differential equations and is appropriate for first-year undergraduates.

10.2 STATEMENT OF THE PROBLEM

In places such as the North of Scotland electricity is generated by hydroelectric means. Despite the high average annual rainfall, the topography of the terrain is such that natural catchment areas tend to be small. In order to make full use of the distributive nature of the catchment areas storage dams are arranged at various levels in the hills and operate by gravity. As well as providing a convenient local storage, this network of dams is used to supply water to the lowest dam, which supplies the flow to drive the turbines and so generates the electric power. An important requirement is to maintain the water level in the lowest dam at an appropriate height to ensure that electric generation demand, which is continuously changing with time, is met. Obviously, because of the remoteness of some of the supply dams, it is desirable that the water supply to the lowest dam be controlled automatically and that the mode of control is capable of responding very quickly to demand changes. The controller should also be capable of taking account of disturbances such as direct inflow to the lowest dam from uncontrolled sources, for example, sudden heavy rainfall or a subsidiary stream.

In order to clarify our ideas regarding the mathematical problems involved consider the two-vessel system in cascade, as shown in Fig. 10.1 where each vessel is assumed to have a uniform cross-sectional area.

Fig. 10.1   Two vessels in cascade

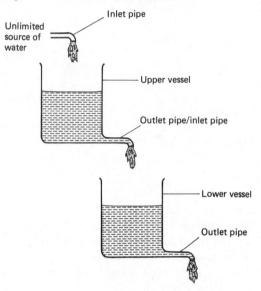

Set up a mathematical model of this two-vessel system and discuss how you would control the inflow to the upper vessel in order to maintain the output flow of the lower vessel at its desired demand value. The amount of water flowing through the inlet and outlet pipes may be controlled by introducing valves. You are required to control the inflow of water in such a way that it can take account of changes in the demand. Also, the controlled system must be capable of reacting to unpredictable disturbances such as a leakage, or an additional uncontrolled inflow, in one, or both, of the vessels.

10.3   ESSENTIAL VARIABLES

At this stage let the students have an opportunity to consider what are the essential variables describing the situation. Also, they should be asked to identify the problem in terms of these variables.

Discussion should lead to the conclusion that the essential variables are:

(a)   the flow rates in and out of each vessel; and

(b)   water level in each vessel.

Introducing these variables into Fig. 10.1 leads to Fig. 10.2. Here, the flow rates, measured for instance in $m^3 s^{-1}$, are denoted by the letter q, which is fairly conventional in mathematics.

Fig. 10.2   Essential variables

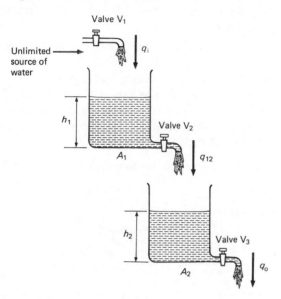

In particular,

$q_i(t)$ denotes the input flow to the upper vessel (hence the suffix i);

$q_{12}(t)$ denotes the flow from the upper to the lower vessel (the suffix 12 denoting flow from vessel 1 to vessel 2);

and $q_0(t)$ denotes the output flow from the lower vessel (hence the suffix 0);

where the term flow is used to represent the rate of flow. The water levels in the upper and lower vessels are denoted by $h_1(t)$ and $h_2(t)$ respectively, both measured in metres (m). The uniform cross-sectional areas of the vessels are denoted by A_1 and A_2 respectively, both measured in m^2. Valves V_1, V_2 and V_3 are introduced in order that the flow rates may be varied.

In terms of the variables introduced the objective is to control the input flow $q_i(t)$, to the upper vessel, in order that the output flow $q_0(t)$, of the lower vessel, continuously attains its desired demand value.

10.4 RELATIONSHIP BETWEEN VARIABLES

The next stage in the model formulation process is to get the students to discuss how the essential variables, as listed above, are interrelated. Points that should be noted are as follows:

(a) Assuming an unlimited source of water the input flow rate $q_i(t)$ to the upper vessel depends only on the setting of valve V_1.

(b) The water level in each vessel will depend on both the input and output flows to that vessel.

(c) The output flow from a particular vessel will depend on several factors:
   (i) the valve settings on the appropriate outlet pipe;
   (ii) the water level in the vessel; the higher the water level the greater the pressure provided to push the water out through the outlet pipe. Consequently, as the water level increases so will the rate of outward flow increase.

Diagrammatically the interrelationship between the variables is illustrated in Fig. 10.3.

Fig. 10.3   Interrelationship between variables

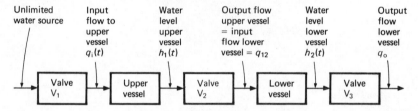

The next problem to be considered is that of determining a mathematical formulation for the interaction between the variables. Basically, two relationships are required, viz:

(a) for a particular vessel the relationship between the water level and the input and output flows (here the water level represents the dependent variable whilst the input and output flows represent the independent variables);

(b) for a particular vessel the relationship between the output flow and the water level (here the output flow represents the dependent variable whilst the water level is the independent variable).

At this stage give the students time to consider these requirements. Suggest they consider a single vessel in isolation as depicted in Fig. 10.4. Here the uniform cross-sectional area is denoted by $A\,\mathrm{m^2}$, the input and output flows by $q_i(t)$, $q_0(t)\,\mathrm{m^3\,s^{-1}}$, respectively, and the water level by $h(t)\,\mathrm{m}$.

Fig. 10.4 Single vessel case

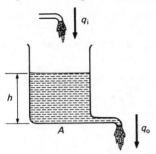

An approach to answering requirement (a) is to make use of the concept of continuity, namely, that the increase in the volume of water in the vessel over a finite interval of time is equal to the volume of water entering minus the volume of water leaving the vessel over the same interval of time. In other words

Rate of increase of volume of water

$$= \text{Rate at which water is entering the vessel}$$

$$- \text{Rate at which water is leaving the vessel}$$

$$= q_i(t) - q_0(t).$$

Now,

Volume of water in the vessel = Cross sectional area $\times$ Water level

$$= Ah(t).$$

Thus,

Rate of increase of volume of water in the vessel $= \dfrac{d}{dt}[Ah(t)]$

$$= A\frac{dh(t)}{dt}.$$

Hence, from above,

$$A\frac{dh(t)}{dt} = q_i(t) - q_0(t) \qquad [10.1]$$

which is a mathematical relationship meeting requirement (a).

Next consider requirement (b). In this case the method of approach is not so obvious. As mentioned earlier the rate of outward flow will increase as the water-level increases, so that the outward flow $q_0(t)$ is an increasing function of the water level $h(t)$. As a first attempt the simplest relationship is assumed, namely that the flow varies in direct proportion to the level so that

$$q_0(t) = \lambda h(t) \hspace{4cm} [10.2]$$

where λ is a constant.

To examine the validity of such an assumption you may choose to proceed in various ways depending on the current mathematical knowledge of the students to whom the case study is directed.

You may wish to ask the students to go away and carry out a practical experiment.[1] This could involve using a set up as shown in Fig. 10.5.

Fig. 10.5 Experimental set-up

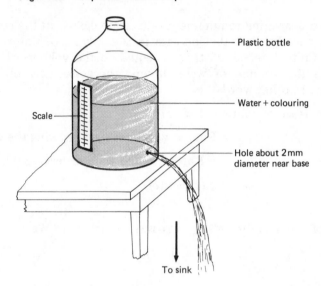

The variation in water level with time is determined experimentally and the results compared to those predicted by the mathematical model. In this case there is no input of water so that $q_i(t) = 0$. Substituting [10.2] in [10.1] gives

$$A\frac{dh(t)}{dt} = -\lambda h(t)$$

which is readily solved to give

$$h(t) = h(0)\exp\left(-\frac{\lambda}{A}t\right)$$

where $h(0)$ is the level of the water in the bottle when the experiment is commenced. Taking natural logarithms this gives

$$\ln h(t) = \ln h(0) - \frac{\lambda}{A}t$$

so that the mathematical model predicts a linear relationship between $\ln h(t)$ and t. Experimental results, however, will predict a relationship of the form shown in Fig. 10.6. It is noted that for large values of $h(t)$ the model appears effective but, in general, the relationship between $h(t)$ and $q_0(t)$ is nonlinear.

At this stage you may ask the students to analyse the experimental data further in order to obtain an empirical relation. Alternatively, you may ask the question 'Is there a relationship between flow rate and water level based on physical principles?' It is likely that the existence of such a principle, known as Torricelli's theorem,[2] will have to be pointed out to the students. The theorem is stated in Appendix 10.1.

Fig. 10.6   Experimental results

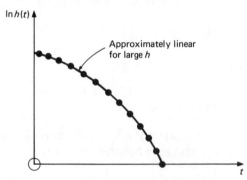

For the dam problem we are concerned with the values of $h(t)$ will always be large and, further, the variations in the water level will be small in comparison with the actual water level. In such cases a linear relationship of the form [10.2] appears to be appropriate; a deduction which is confirmed in Appendix 10.1.

In summary, it is concluded that, for a single vessel, provided the variation in water level is small in comparison with the actual level then the relationships between the essential variables may be represented by Equations [10.1] and [10.2]. Substituting [10.2] in [10.1] gives the relationship between the water level $h(t)$ and input flow $q_i(t)$ as

$$A \frac{\mathrm{d}h(t)}{\mathrm{d}t} + \lambda h(t) = q_i(t) \qquad [10.3]$$

10.5   MODELLING THE TWO-VESSELS ARRANGEMENT

Returning to the original problem, depicted in Fig. 10.2, the overall objective is to control the input $q_i(t)$ in order to provide the appropriate outflow $q_0(t)$ to meet the demand. However, before considering the question of control it is necessary to obtain a model relating $q_0(t)$ to $q_i(t)$.

For the upper vessel the equations corresponding to Equations [10.1] and [10.2] are

$$A_1 \frac{dh_1(t)}{dt} = q_i(t) - q_{12}(t) \qquad [10.4]$$

and

$$q_{12}(t) = \alpha_1 h_1(t) \qquad [10.5]$$

where α_1 is a constant, dimensions $m^2 s^{-1}$, which may be varied by varying the setting of valve V_2. Eliminating $q_{12}(t)$ gives the relationship between the water level $h_1(t)$ and the input flow $q_i(t)$ as

$$A_1 \frac{dh_1(t)}{dt} + \alpha_1 h_1(t) = q_i(t). \qquad [10.6]$$

Similarly, for the lower vessel

$$A_2 \frac{dh_2(t)}{dt} + \alpha_2 h_2(t) = q_{12}(t) \qquad [10.7]$$

where α_2 is a constant, dimensions $m^2 s^{-1}$, which may be varied by varying the setting of valve V_3, and satisfies the equation

$$q_0(t) = \alpha_2 h_2(t). \qquad [10.8]$$

Equations [10.6] and [10.7] may now be used to eliminate the variables $q_{12}(t)$ and $h_1(t)$. This elimination may be carried out in various ways, a possible approach being as follows.

Writing $D \equiv d/dt$ Equation [10.7] becomes

$$(A_2 D + \alpha_2) h_2(t) = q_{12}(t)$$

$$= \alpha_1 h_1(t) \quad \text{using Equation [10.5]}.$$

Operating on both sides with $(A_1 D + \alpha_1)$ gives

$$(A_1 D + \alpha_1)(A_2 D + \alpha_2) h_2(t) = \alpha_1 (A_1 D + \alpha_1) h_1(t) \qquad [10.9]$$

$$= \alpha_1 q_i(t) \quad \text{using Equation [10.6]}.$$

Thus, a mathematical model relating the water level $h_2(t)$ of the lower vessel, to the input flow $q_i(t)$, of the upper vessel, is the second-order differential equation

$$\frac{d^2 h_2(t)}{dt^2} + \frac{(\alpha_1 A_2 + \alpha_2 A_1)}{A_1 A_2} \frac{dh_2(t)}{dt} + \frac{\alpha_1 \alpha_2 h_2(t)}{A_1 A_2} = \frac{\alpha_1}{A_1 A_2} q_i(t). \qquad [10.10]$$

The relationship between the output flow $q_0(t)$ and $h_2(t)$ is then given by Equation [10.8]. Since this is taken to be a linear relationship it follows that the problem of providing a required demand output $q_0(t)$ is equivalent to that of providing the corresponding appropriate water level $h_2(t)$.

10.6 MODEL INTERPRETATION

Equation [10.10] may be written in the general form

$$\frac{d^2h(t)}{dt^2} + 2\xi\omega_n \frac{dh(t)}{dt} + \omega^2_n h(t) = \frac{\alpha_1}{A_1 A_2} q_i(t) \qquad [10.11]$$

discussed in Appendix 10.2. The dimensionless damping ratio ξ, and natural frequency ω_n being given by

$$2\xi = \frac{\alpha_1 A_2 + \alpha_2 A_1}{\sqrt{\alpha_1 \alpha_2 A_1 A_2}} \quad \text{and} \quad \omega_n{}^2 = \frac{\alpha_1 \alpha_2}{A_1 A_2}.$$

When introducing parameters such as ξ and ω_n it is advisable to have a dimension check. Since α_1 and α_2 have dimensions $m^2 s^{-1}$ and A_1, A_2 have dimensions m the check is satisfactory.

In this case

$$\xi^2 - 1 = (\alpha_1 A_2 - \alpha_2 A_1)^2/(4\alpha_1 \alpha_2 A_1 A_2) > 0.$$

Thus, the system is overdamped and an oscillatory response is not possible. (**Note:** The fact that the corresponding characteristic equation has distinct real roots is clearly indicated in Equation [10.9].) As is seen in Appendix 10.2, a consequence of this is that the system will respond slowly to changes in demand. The best performance is achieved by taking $\xi = 1$.   Since the cross-sectional areas A_1 and A_2 are fixed the settings of the outlet valves V_2 and V_3 (which determine the values of α_1 and α_2) may be chosen so that $\xi = 1$.

Having chosen the values of α_1 and α_2 to meet this specification we may argue that all that is needed is to solve Equation [10.10] for various constant values of $q_i(t)$. In this way a table of values of $h_2(t)$ versus $q_i(t)$ may be drawn up. This may then be used to calibrate the inlet valve V_1 so that particular valve settings correspond to particular demand water levels $h_2(t)$ of the lower vessel (which in turn correspond to particular demand outflow $q_0(t)$). As the demand changes, the valve settings may be changed accordingly, possibly by remote control.

It appears that a suitable model has been found for the problem. Does it meet all the specified requirements? Apart from responding slowly to changes in demand, the model has some other shortcomings. For example, it cannot take account of factors such as the following.

(a) If either of the vessels develops a leak then the required demand will not be met using the setting for valve V_1 determined by the model. Similarly, if either of the vessels is subject to an additional inflow, other than from the controlled source (e.g. in the case of the dams the effect of rain or inflow from small tributaries) then the response will not be as predicted by the model.

(b) The model takes no account of the efficiency of valves V_1, V_2 and V_3. It is assumed that they function properly all the time.

In summary, the model is such that it has no way of monitoring its own behaviour and is, therefore, not capable of reacting to disturbances such as (a) and (b).

10.7 MODEL ENHANCEMENT

In order to improve the model what is needed is a means of introducing a self-monitoring capability.

At this stage give the students an opportunity to go away and consider means of achieving such an enhancement.

A possible way of proceeding is now considered. Suppose that at a particular instant in time the desired water level of the lower vessel is $h_D(t)$. Then, introducing the concept of self-monitoring, Fig. 10.3 may be replaced by Fig. 10.7.

Fig. 10.7   Introduction to self-monitoring

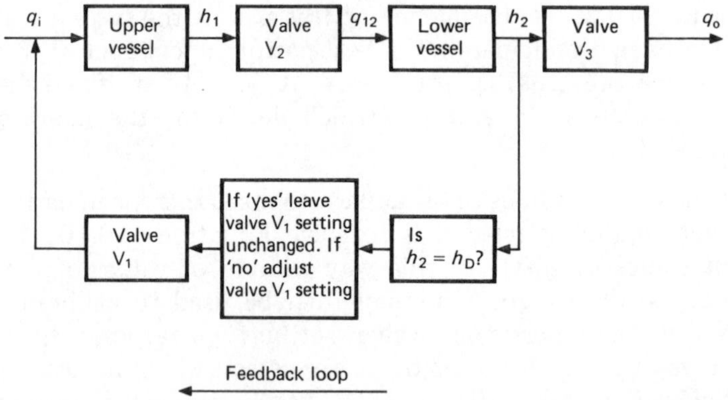

A system, such as that depicted in Fig. 10.7 which has the ability to monitor its own behaviour is called a 'feedback control system'.[3] The updated version of Fig. 10.2 is shown in Fig. 10.8. The level sensor measures the water level $h_2(t)$; the comparator, or error detector, device then compares this value with the desired value $h_D(t)$ and generates an error signal $[h_D(t) - h_2(t)]$. This error signal then actuates the controller device which operates the valve V_1.

Fig. 10.8 Feedback system

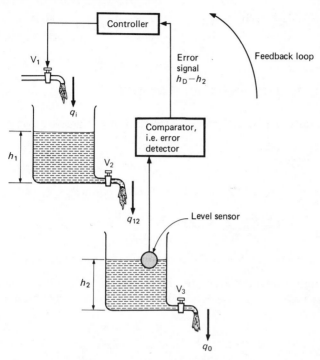

What is the mathematical model representing this revised configuration? Again give the students an opportunity to develop this for themselves.

The only difference between the system of Fig. 10.8, and that of Fig. 10.2 is that the inflow $q_i(t)$ to the upper vessel is now a function of the error signal $[h_D(t) - h_2(t)]$, the error signal being the difference between the desired and the actual water level of the lower vessel. Obviously, there are various functional relationships one may adopt. The simplest is to assume that the valve V_1 is operated in such a way that the inflow $q_i(t)$ varies directly in proportion to $[h_D(t) - h_2(t)]$, i.e.

$$q_i(t) = K[h_D(t) - h_2(t)] \qquad\qquad [10.12]$$

where K is a constant having dimension $m^2\,s^{-1}$.

At this stage the students should be asked to go away and incorporate Equation [10.12] into the previous model and discuss its resulting behaviour.

From Equation [10.10] a mathematical model for the revised set-up is the second-order differential equation

$$\frac{d^2h_2(t)}{dt^2} + \frac{(\alpha_1 A_2 + \alpha_2 A_1)}{A_1 A_2} \frac{dh_2(t)}{dt} + \frac{\alpha_1 \alpha_2 h_2(t)}{A_1 A_2} = \frac{\alpha_1 K}{A_1 A_2} [h_D(t) - h_2(t)]$$

or, equivalently,

$$\frac{d^2h_2(t)}{dt^2} + \frac{(\alpha_1 A_2 + \alpha_2 A_1)}{A_1 A_2} \frac{dh_2(t)}{dt} + \frac{(\alpha_1 \alpha_2 + \alpha_1 K)}{A_1 A_2} h_2(t) = \frac{\alpha_1 K}{A_1 A_2} h_D(t).$$

[10.13]

10.8 INTERPRETATION OF ENHANCED MODEL

Differential equation [10.13] may be written in the general form

$$\frac{d^2h_2(t)}{dt^2} + 2\xi\omega_n \frac{dh_2(t)}{dt} + \omega^2_n h_2(t) = \frac{\alpha_1 K}{A_1 A_2} h_D(t)$$

where the new damping ratio ξ and natural frequency ω_n are given by

$$2\xi = \frac{\alpha_1 A_2 + \alpha_2 A_1}{\sqrt{A_1 A_2}\sqrt{\alpha_1 \alpha_2 + \alpha_1 K}}, \quad \omega^2_n = \frac{\alpha_1(\alpha_2 + K)}{A_1 A_2}.$$

[10.14]

(**Note:** Remember to carry out a dimensional check at this stage.)

For a particular constant demand level $h_D(t) = H$ the system will behave as discussed in Appendix 10.2, with the steady-stage response being given by

$$[h_2(t)]_{\text{steady state}} = \frac{\alpha_1 K}{A_1 A_2} \frac{A_1 A_2}{\alpha_1(\alpha_2 + K)} H$$

$$= \left(\frac{1}{1 + \alpha_2/K}\right) H.$$

[10.15]

Thus, the water level $h_2(t)$ of the lower vessel will never actually reach the desired value H; it will fall short by a factor $[1/(1 + \alpha_2/K)]$. The fact that the amount of shortfall decreases as the value of K increases leads to the suggestion that a very high value of K should be used.

Before reaching such a conclusion the other effects of increasing the value of K must first be considered.

In this case

$$\xi^2 - 1 = \frac{(\alpha_1 A_2 - \alpha_2 A_1)^2 - 4A_1 A_2 \alpha_1 K}{4\alpha_1 \alpha_2 A_1 A_2}$$

which may take positive or negative values. Thus, the introduction of the feedback loop leads to the possibility of a damped oscillatory response ($\xi < 1$) which, as discussed in Appendix 10.2 is a desirable feature. As the value of K increases so the value of ξ decreases leading eventually to a large overshoot, resulting in an increase in the time taken for the system to settle down to its steady-state value, which is an undesirable feature. From Equations [10.14] it is seen that another effect of increasing K is to increase the natural frequency ω_n, which, in practice, is found to be desirable as it helps to eliminate the effects of disturbances. Thus, there are conflicting requirements when choosing an appropriate value for K.

(a) On the one hand it is desirable to choose K as large as possible so as to reduce the amount of shortfall; but

(b) on the other hand large values of K implies less damping thus leading to a long settling-down time.

A compromise must be made and a value of K, together with values of α_1 and α_2, chosen so that the shortfall is sufficiently small to be acceptable whilst the value of ξ is such that $0.6 < \xi < 0.8$, thus ensuring that the settling-down time is not too large.

In conclusion, the updated model has the ability to monitor its own behaviour, and therefore, should be capable of coping with the deficiencies highlighted for the previous model. However, some compromise has to be made to avoid too long a delay in the time taken for the water level to settle down to its steady state level. A consequence of this compromise is that there will always be a shortfall in the actual water level in relation to the desired value. For many practical applications, such as the hydroelectric power generation system, this model could well be acceptable as the amount of shortfall is within the percentage tolerance acceptable.

10.9 FURTHER WORK

1. By incorporating appropriate terms into the mathematical model discuss the effects of disturbances, such as a leakage or uncontrolled additional inflow, in relation to the lower vessel.

2. Update the model in order to eliminate the shortfall factor. Discuss the behaviour of the updated model (**Hint:** What is required is to build up the error signal over an interval of time so that it becomes large enough to activate the controller.)

3. In practice the demand $h_D(t)$ is likely to vary periodically over a 24 hour period. Discuss various forms of variation and how the model responds to such demands.

4. There will be a time delay between the adjustment of valve V_1 and q_i reaching its desired value. Discuss the incorporation of such an effect into the model.

10.10 ALTERNATIVE EXERCISE

Rather than considering the two vessels in cascade consider the two vessels in parallel as in Fig. 10.9. The objective is to control the inflow q_i such that the outflow q_0 attains a desired value. Such an arrangement of vessels is used in the chemical processing industry.

Fig. 10.9   Vessels in parallel

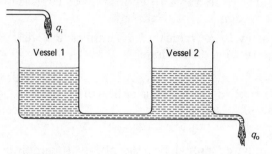

APPENDIX 10.1

FLOW RATE PROPORTIONAL TO WATER LEVEL

Fig. 10.10

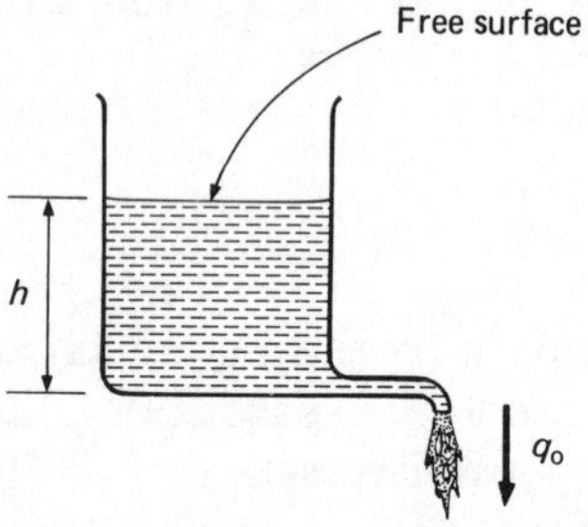

Based on physical principles the relationship between the flow rate q_0 and the water level h is given by Torricelli's theorem. The theorem states:

When a non-viscous fluid emerges in a jet from a hole in a vessel, where the hole is much smaller than the area of the free surface of the fluid in the vessel, the velocity v, $m\,s^{-1}$, of the jet at the hole is

$$v = \sqrt{2gh}$$

where $h\,m$, is the level of the fluid in the vessel above the hole and g is the gravity constant having value $9.81\,m\,s^{-2}$.

The rate of outflow $q_0(t)$ is then related to the velocity v by

$$q_0 = \mu v$$

where the constant μ is the area of the hole in m^2. Thus, the relationship between q_0 and h is

$$q_0 = c\sqrt{h} \qquad [10.16]$$

where c is a constant having dimension $m^{5/2}\,s^{-1}$.

[If the experiment outlined in Section 10.3 has been carried out use the experimental results to validate Torricelli's theorem. Substituting Equation [10.16] in Equation [10.1], with $q_i(t) = 0$, gives

$$A\frac{dh(t)}{dt} = -c\sqrt{h}$$

which may be solved to give

$$2\sqrt{h(t)} = 2\sqrt{h(0)} - \frac{c}{A}t$$

thus predicting a linear relationship between $\sqrt{h(t)}$ and t. Confirm that this agrees with the $\ln h(t)$ against t graph obtained from the experimental results.]

From [i]

$$q_0(h) = c\sqrt{h} = c[H + (h - H)]^{1/2} \quad \text{for any } H$$

$$= cH^{1/2}\left[1 + \frac{h - H}{2H}\right], \quad \text{provided } h - H \text{ is small}$$

$$= q_0(H) + \frac{c}{2\sqrt{H}}(h - H).$$

That is, provided $(h - H)$ is small

$$q_0(h) \simeq q_0(H) + \alpha(h - H) \qquad [10.17]$$

where the constant $\alpha = c/2\sqrt{H}$.

The significance of each term in Equation [10.17] may be seen from Fig. [10.11].

For large values of h the situation is illustrated in Fig. 10.12. It is seen that as the value of h increases $q_0(H) - \alpha H$ becomes smaller in comparison to αh and

$$q_0(h) \simeq \alpha h. \qquad [10.18]$$

Thus provided, as in the reservoir case, the variations in water level are small in comparison to the actual water level, which is large, then result [10.18] is valid.

Fig. 10.11

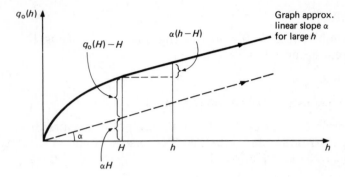

Fig. 10.12

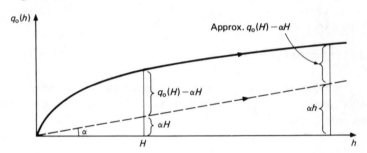

APPENDIX 10.2

RESPONSE OF A SECOND-ORDER SYSTEM TO A CONSTANT INPUT

Consider a second-order system which may be represented mathematically by a second-order differential equation with constant coefficients. Such a differential equation may be written in the general form

$$\frac{d^2y}{dt^2} + 2\xi\omega_n \frac{dy}{dt} + \omega_n^2 y = x(t) \tag{10.19}$$

where $y(t)$ is the system output and $x(t)$ its input. ξ is a dimensionless parameter, called the damping ratio, and ω_n a parameter, called the natural frequency, having units of 1/seconds. For practical systems $\xi > 0$ and $\omega_n > 0$.

The general solution of Equation [10.19] is the sum of any particular solution and the complementary function,[4] the complementary function being the general solution of the zero input system

$$\frac{d^2y}{dt^2} + 2\xi\omega_n \frac{dy}{dt} + \omega_n^2 y = 0. \tag{10.20}$$

Trying a solution $y = ce^{mt}$ leads to the characteristic equation for Equation [10.20]

$$m^2 + 2\xi\omega_n m + \omega_n^2 = 0$$

having roots

$$m_1, m_2 = -\xi\omega_n \pm \omega_n\sqrt{\xi^2 - 1}. \qquad [10.21]$$

Three cases arise depending on the value of ξ.

Case I, $\xi > 1$ (overdamped case)

In this case the roots [10.21] are real, distinct and both negative.

Thus, the general solution of Equation [10.20] is

$$y(t) = Ae^{m_1 t} + Be^{m_2 t}$$

where A and B are arbitrary constants. Since both m_1 and m_2 are negative it follows that the output $y(t)$ decreases exponentially with time.

Case II, $0 < \xi < 1$ (underdamped case)

In this case the roots [10.21] are complex conjugates,

$$m_1, m_2 = -\xi\omega_n \pm j\omega_n\sqrt{1 - \xi^2}.$$

The corresponding general solution of Equation [10.20] is

$$y(t) = e^{-\xi\omega_n t}[A \cos \omega_n\sqrt{1 - \xi^2}\,t + B \sin \omega_n\sqrt{1 - \xi^2}\,t]$$

where A and B are arbitrary constants. Since both ξ and ω_n are positive it follows that the output $y(t)$ is a damped oscillation.

Case III, $\xi = 1$ (critically damped)

This corresponds to the boundary between $\xi > 1$ and $\xi < 1$. The roots [10.21] are equal and the corresponding general solution of Equation [10.20] is

$$y(t) = (A + Bt)e^{m_1 t}$$

where A and B are arbitrary constants. The output $y(t)$ looks similar to the case $\xi > 1$.

Returning to differential equation [10.19], for the particular case of the input $x(t)$ being a constant R, then a particular solution is

$$y = \frac{R}{\omega_n^2}.$$

This is referred to as the steady-state output of the system and is the value the output approaches as $t \to \infty$. How the output approaches this steady-state value depends on the value of the damping ratio ξ. If $\xi > 1$ then the approach will be exponential whilst it will be damped oscillatory in the case $0 < \xi < 1$. Solutions for various values of ξ are illustrated in Fig. 10.13.

Fig. 10.13

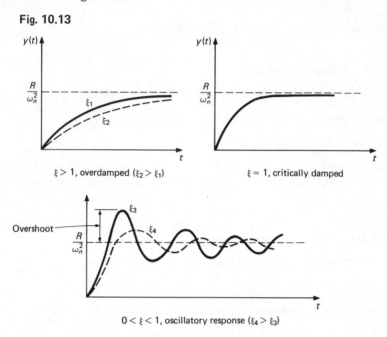

$\xi > 1$, overdamped ($\xi_2 > \xi_1$)

$\xi = 1$, critically damped

$0 < \xi < 1$, oscillatory response ($\xi_4 > \xi_3$)

It is seen that for the case $\xi > 1$ the response is slow. For the case $0 < \xi < 1$ the response is more rapid the smaller ξ; on the other hand the effect of small ξ is to produce a large overshoot and hence increase the time taken for the system to settle down.

In practice a control system should possess a fast action and a small overshoot which provide conflicting demands on ξ. For the type of system being discussed in this case study a value of ξ in the range $0.6 < \xi < 0.8$ is found to be acceptable.

REFERENCES

1. *The Open University Mathematical Foundation Course M101*, Block V, Mathematical Modelling. Unit 2, Time Dependent Models (Milton Keynes, Open University, 1979)
2. L.M. Milne-Thompson *Theoretical Hydrodynamics* (London, MacMillan, 1955)
3. Di Stefano, Stubber and Williams, *Feedback and Control Systems*, Schaum's Outline Series (New York, McGraw Hill, 1967)
4. Any textbook on differential equations, e.g. K.A. Stroud *Engineering Mathematics* (London, Macmillan, 1970)

11.

Speed-Wobble in Motorcycles

K.H. Oke

11.1   PRE-REQUISITES

Elementary mechanics of a rigid body, and SI units. Acquaintance with analytical and numerical methods of solving ordinary differential equations.

11.2   STATEMENT OF THE PROBLEM

Many of us have experienced wobbling of the front wheel, felt through the handle-bar, of a motorcycle (or of an ordinary bicycle) when travelling at certain speeds. What causes this wobble, or oscillation, of the steered wheel?

The wobbling phenomenon is not confined to motorcycles and bicycles but it is also known to occur in the front wheels of cars, supermarket or tea trolleys, and in aircraft nose-wheels. These wide-ranging situations all have something in common, namely that the steered wheel is designed as a castor. A castor is defined as a steered rolling wheel, whose point of contact with the ground lies behind the point of intersection of the steering axis and the ground. Fig. 11.1 illustrates the basic configuration of two typical castors (not drawn to scale).

Note that the rake angle of a supermarket trolley castor is 0°, that is, the steering axis is vertical. Clearly a number of other important features have to be considered in any attempt to understand how castors oscillate.

Fig. 11.1 Typical castors

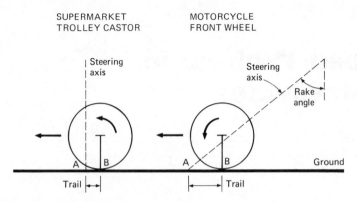

For motorcycles, the tyre of the steered wheel and also the suspension in the front forks will obviously affect steering stability. Consequently, there will not be hard (point) contact between the wheel and ground. The area of the 'contact patch' depends on tyre pressure, forward speed of the motorcycle, and whether the cycle is banking on a bend.[1] The flexibility of the tyre also permits lateral movement of the wheel without slipping.

It has been found in practice[1] that the front wheel, even in the 'wheel locked' case (i.e. brakes jammed on hard), can in the case of motorcycles oscillate with a frequency somewhere in the range of 6–8 Hz. In stable cases (which hopefully form the vast majority!), these oscillations rapidly decay to zero.

The problem, then, is to formulate a model which explains some or all of these observations.

11.2.1 DATA

Typical motorcycle values, for front wheel oscillations of 6–8 Hz are:

Speed of motorcycle $= 30\,\mathrm{ms}^{-1}$

Moment of inertia of wheel about steering axis $= 0.27\,\mathrm{kg\,m}^2$

Trail $= 0.12\,\mathrm{m}$

Coefficient of friction between wheel and ground $= 1$

Normal reaction between wheel and ground $= 700\,\mathrm{N}$

Angular velocity of front wheel about steering axis $= \pm\,12\,\mathrm{rad\,s}^{-1}$ when wheel not turned, i.e. when turning angle is 0.

11.3 HOW MODELLING EXERCISE MIGHT BE PRESENTED TO STUDENTS

This problem is considered to be fairly difficult both in the formulation stage and in handling what could be an unfamiliar piece of mathematics to many undergraduates. If students have a generally good analytical and numerical mathematics background, as would normally be expected, say, in second or third year degree courses in mathematics/science/engineering, then they ought to be able to make good progress with the 'statement of the problem' with only a few hints from the lecturer. This further assumes that students have already some experience in mathematical modelling.

For those students who have little or no experience in modelling, it is suggested that the following approach be attempted.

With interactive teaching, allow approximately two $1\frac{1}{2}$ hour sessions with the class in formulating a simple model of the original motorcycle problem as described below.

(a) Present 'Statement of the Problem' and discuss with class. Include brief refresher on such things as SI units and concepts of friction, inertia, equations of motion.

(b) Invite students to identify variables and to make assumptions in formulation stage. List variables and assumptions, as they are made, on the blackboard.

(c) Ask students if they think that there are enough variables and assumptions to enable them to attempt a solution (or solutions). Do students think that this will not be known until a solution is actually attempted? At what stage (if at all) is a vertical-axis castor considered? If the latter is not thought of by the students, the lecturer might introduce this simplification at this stage and discuss its possible use in understanding the original motorcycle problem with the class — remembering to allow student's objections as well as other comments.

(d) Ask students to guess outcome — for example, how does the castor oscillate? What confidence have the students in their intuitive understanding of the problem? A simple experiment with a vertical-axis castor with the wheel in contact with a moving belt would be most illuminating at this juncture.

(e) So far, steps (a)–(d) will (depending on the class) need about $1\frac{1}{2}$ hours interactive development. For the second $1\frac{1}{2}$ hours session, it is suggested that students write down the equations (with the minimum help from the lecturer) for cases of vertical-axis castor motion — letting students identify various cases, for example, wheel-locked (with or without side slipping), wheel-rotating (with or without side slipping). At this stage, students should be allowed to work individually or in small groups. At the end of the session students should have reached a point of development comparable to that given in expressions [11.2] and/or [11.5].

After the two $1\frac{1}{2}$ hour sessions ask students to work individually in solving the equations of motion — it is important to encourage numerical solutions and graphical interpretation. It should be appreciated that probably only the better students will progress in a comparable way to the solution and interpretation of the solution as indicated in the following development. Follow-up exercises and extensions to the model are suggested for students, set either as homework or coursework.

A particular approach in tackling the problem now follows, but it should be realised that this is only one particular development — students/lecturer may well develop different models.

11.4 FORMULATION OF THE PROBLEM

It is fairly clear that if the front wheel of a motorcycle is going to oscillate, then it will start to do so the moment the plane of the wheel is perturbed (by steering, or perhaps by a small object on the road surface) from the forward direction of motion. Once it is oscillating we require to know which of the key design features above affect the frequency. What are the conditions for stability — that is, for the oscillations to decay rapidly? In practice, it seems desirable for stability that the oscillations should decay in at most one or two seconds.

For convenience, we list the design features already identified and which appear to have some bearing on the problem.

1. Spatial relationship of steering axis to wheel determined by rake angle and trail distance.
2. Tyre flexibility which allows lateral movement of wheel without slipping.
3. Tyre contact patch area.
4. Suspension in front forks.

Other factors which may affect the problem include wheel-bearing play (allowing further lateral movement), lateral flexibility of front forks (and hence of wheel) and gyroscopic couples introduced because the steering axis is not vertical. We can assume that the wheel has been dynamically balanced by a tyre specialist.

We have, then, what appears to be a very complex situation. It seems necessary to make a number of simplifying assumptions in order to define a conceptually easier problem. It is to be hoped that the 'easier problem' will lead to some manageable mathematics, and that some deductions can be made which will provide useful insights to the original situation of motorcycles.

From what has been discussed so far, it is now becoming clear that it would be easier to consider a supermarket trolley castor, which has a tyreless wheel and a rigid vertical steering axis. Gyroscopic couples can then be ignored as well as tyre flexibility and suspension in the steering axis. As we will now consider a tyreless wheel, we might also assume hard (point) contact with the ground. In practice, however, even with a tyreless wheel, some frictional resistance is felt when the wheel is steered; this is mainly due to resistance at the small, but finite, area of contact between the wheel and the ground. A smaller resistance, which we will ignore, is due to a frictional torque in the steering column bearings.

We are now at the crux of our simplified model. How should we represent the frictional forces at the point of contact of the wheel with the ground? We are already dealing with a much simplified problem and it would seem appropriate to make the following assumptions regarding frictional forces.

(a) Simple Coulomb friction applies, with frictional force μR proportional to normal reaction R, with μ the dynamical coefficient of friction.

(b) This Coulomb friction applies irrespective of the forward speed v of the castor.

(c) Coulomb friction applies both when the wheel is locked and when the wheel is rolling.

(d) The direction of the frictional force, μR, is directly opposite to the direction of motion of the contact point of the castor with the ground.

We will further assume that the forward speed v is constant.

11.5 DEVELOPMENT OF A MODEL

WHEEL-LOCKED CASE

Fig. 11.2 shows the basic geometry of the vertical steering axis castor and Fig. 11.3 shows the velocity components, as well as the frictional force components, at the point of contact B. Note that since the wheel is locked, the velocity of B is the same as the steering axis A plus the velocity $L\dot{\theta}$ perpendicular to AB due to angular rotation $\dot{\theta}$ of AB. The resultant velocity of B is therefore given by

$$V = \sqrt{v^2 + (L\dot{\theta})^2 + 2Lv\dot{\theta}\sin\theta} \qquad [11.1]$$

Fig. 11.2 Plan view of vertical steering axis castor

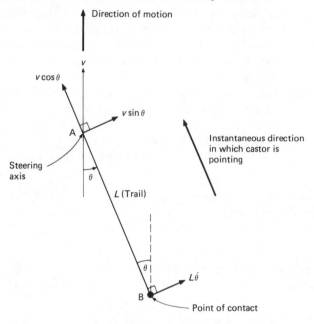

Fig. 11.3 Components of velocity and components of frictional force at point of contact (wheel locked)

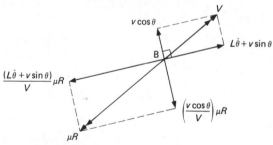

The frictional force μR is in the opposite direction to V (assumption (d)), and has the components shown; the values of these components are seen on inspection in view of the similar rectangles of vectors involved.

Taking moments about A, we obtain the following equation of motion

$$I\ddot{\theta} = -\frac{(L\dot{\theta} + v\sin\theta)L}{V}\mu R \qquad [11.2]$$

where I is the moment of inertia of the wheel and its support about A. For small θ, that is, for oscillations of small amplitude, we have from Equations [11.1] and [11.2]

$$I\ddot{\theta} + \frac{\mu R L^2 \dot{\theta}}{\sqrt{v^2 + (L\dot{\theta})^2 + 2Lv\dot{\theta}\theta}} + \frac{\mu R L v\theta}{\sqrt{v^2 + (L\dot{\theta})^2 + 2Lv\dot{\theta}\theta}} = 0. \qquad [11.3]$$

For motorcycles, $v^2 \gg (L\dot{\theta})^2 + 2Lv\dot{\theta}\theta$, as can be seen from taking typical values: $v = 30\,\mathrm{m\,s^{-1}}$, $L = 0.12\,\mathrm{m}$, $\dot{\theta} = 12\,\mathrm{rad\,s^{-1}}$ and $\theta = 0.1$ rad.

Simplifying Equation [11.3] gives

$$I\ddot{\theta} + \left(\frac{\mu R L^2}{v}\right)\dot{\theta} + (\mu R L)\theta = 0. \qquad [11.4]$$

This is the form for damped simple harmonic motion.

ROTATING WHEEL CASE

Since in this case (the more usual one) the wheel is rolling, the resultant velocity of the point of contact B along the direction of AB will be zero. Consequently, the velocity of B is $L\dot{\theta} + v\sin\theta$ in a direction perpendicular to AB. The frictional force μR is directly opposite in direction. Referring to Figs. 11.2 and 11.4 and taking moments about A provides the following equation of motion:

$$I\ddot{\theta} = -\mu R L \,\mathrm{sgn}\,(L\dot{\theta} + v\sin\theta) \qquad [11.5]$$

where the sgn (sign) function is defined as follows:

$$\mathrm{sgn}\,(x) = \begin{cases} 1 & \text{if } x > 0 \\ 0 & \text{if } x = 0 \\ -1 & \text{if } x < 0. \end{cases}$$

Fig. 11.4 Components of velocity and frictional force at point of contact (rotating wheel)

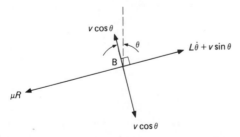

In other words, the frictional force μR is always in a direction directly opposite that of the velocity of B, $L\dot{\theta} + v\sin\theta$. Note that it was not necessary to include the sgn function in Equation [11.2], the wheel-locked case, since the frictional force

$$\left(\frac{L\dot\theta + v\sin\theta}{V}\right)\mu R$$

will inevitably change sign as θ and $\dot\theta$ vary.

For small amplitude oscillations Equation [11.5] may be written

$$I\ddot\theta = -\mu RL\epsilon \qquad\qquad\qquad [11.6]$$

where

$$\epsilon = \text{sgn}\,(L\dot\theta + v\theta). \qquad\qquad\qquad [11.7]$$

It is not clear on first inspection whether or not the second-order non-linear differential equation [11.6] will have a damped oscillatory solution. Although it is clear that the solution will be oscillatory, we do not know whether the period is constant over all complete oscillations.

11.6   ANALYSIS OF MODEL

We now attempt a numerical solution of Equation [11.6], the rotating wheel case, in order to get a better understanding of the nature of the oscillations.

Assuming initial conditions

$$\theta = \theta_0, \quad \dot\theta = \omega_0, \quad \text{when} \quad t = t_0, \qquad\qquad [11.8]$$

integrating Equation [11.6] provides

$$\dot\theta = -\frac{\mu RL\epsilon}{I}\,(t-t_0) + \omega_0 \qquad\qquad\qquad [11.9]$$

and integrating again, we obtain

$$\theta = -\frac{\mu RL\epsilon}{2I}(t-t_0)^2 + \omega_0(t-t_0) + \theta_0. \qquad\qquad [11.10]$$

Note that Equations [11.9] and [11.10] are valid only for those values of t $(\geqslant t_0)$ for which ϵ has the same value as ϵ_0, which from Equation [11.7] is given by

$$\epsilon_0 = \text{sgn}\,(L\omega_0 + v\theta_0). \qquad\qquad\qquad [11.11]$$

This is easily seen as follows. Suppose, for example, that $\epsilon_0 = +1$ and that $t = t_1$ the first time that $\epsilon = -1 = \epsilon_1$ (say), Let $\theta = \theta_1$ and $\dot\theta = \omega_1$, when $t = t_1$. Then if we treat these values as new initial conditions, Equation [11.9] becomes

$$\theta = + \frac{\mu R L}{I}(t-t_1) + \omega_1 \tag{11.12}$$

for $t_1 \leqslant t < t_2$, where t_2 is the time that $\epsilon = +1 = \epsilon_2$ (say).

Fig. 11.5 Graph showing variation of θ with t where $\epsilon_0 = +1$

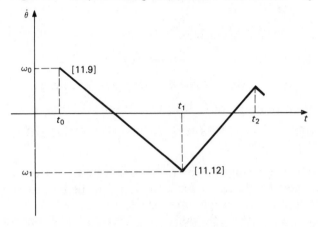

Fig. 11.5 illustrates Equations [11.9] and [11.12] with $\epsilon_0 = +1$; with $\epsilon_0 = -1$, the graph would be reflected in the t-axis. Similarly for θ given by Equation [11.10]. Note that although $\ddot{\theta}$ is discontinuous, both $\dot{\theta}$ and θ are continuous functions of time.

The following scheme for obtaining $\dot{\theta}$ and θ suggests itself:

$$\dot{\theta} = -\frac{\mu R L \epsilon_r}{I}(t-t_r) + \omega_r$$

$$\theta = -\frac{\mu R L \epsilon_r}{2I}(t-t_r)^2 + \omega_r(t-t_r) + \theta_r \tag{11.13}$$

$$\epsilon_{r+1} = -\epsilon_r = -\operatorname{sgn}(L\omega_r + v\theta_r)$$

$$t_r \leqslant t < t_{r+1}; \quad \theta = \theta_r, \quad \dot{\theta} = \omega_r \quad \text{when} \quad t = t_r$$

$$r = 0, 1, 2, \dots.$$

Note that ϵ changes its value the instant $(L\dot{\theta} + v\theta)$ equals zero. So, in using the scheme [11.13], one needs to test $(L\dot{\theta} + v\theta)$ against some small pre-set value, $\alpha > 0$, at each value of t (which is gradually incremented). The instant that $|L\dot{\theta} + v\theta| < \alpha$, t is set equal to t_{r+1} and the calculation continues with $\epsilon_{r+1} = -\epsilon_r$, $\dot{\theta} = \omega_{r+1}$ and $\theta = \theta_{r+1}$ as new initial conditions.

Using typical motorcycle values of $\mu = 1$, $R = 700 \, \text{N}$, $L = 0.12 \, \text{m}$, $I = 0.27 \, \text{kg m}^2$, $v = 30 \, \text{m s}^{-1}$ and initial conditions $\theta = 0$, $\dot{\theta} = -12$ rad s^{-1} the scheme becomes

$$\dot{\theta} = -311.11\,\epsilon_r(t-t_r) + \omega_r$$

$$\theta = -155.56\,\epsilon_r(t-t_r)^2 + \omega_r(t-t_r) + \theta_r \qquad\qquad [11.14]$$

$$\epsilon_r = \mathrm{sgn}\,(0.12\omega_r + 30\theta_r).$$

Thus, with $\theta_0 = 0$, $\omega_0 = -12\,\mathrm{rad\,s^{-1}}$, $\epsilon_0 = \mathrm{sgn}\,[0.12(-12) + 30(0)]$, i.e. $\epsilon_0 = \mathrm{sgn}\,(-1.44) = -1$.

With $t = 0.05\,\mathrm{s}$, the values are

$$\dot{\theta} = +311.11(0.05-0) - 12 = 3.555\,\mathrm{rad\,s^{-1}}$$

$$\theta = +155.56(0.05-0)^2 - 12(0.05-0) = -0.2111\,\mathrm{rad}$$

$$\epsilon = \mathrm{sgn}\,[0.12(3.555) + 30(-0.2111)] = \mathrm{sgn}\,(-5.903) = -1.$$

With $t = 0.075\,\mathrm{s}$, $\dot{\theta} = 11.33\,\mathrm{rad\,s^{-1}}$, $\theta = -0.02475\,\mathrm{rad}$, $\epsilon = \mathrm{sgn}\,(0.6171) = +1$.

Thus $(L\dot{\theta} + v\theta)$ has changed sign at a time t where $0.05 < t < 0.075$, and so the last values calculated for $\dot{\theta}$ and θ will be incorrect. We now try simply 'chopping' the time interval to locate that t $(= t_1)$ for which $|L\dot{\theta} + v\theta|$ is small (< 0.0005 say). Table 11.1 shows the values obtained.

Table 11.1 Table of values with $\epsilon_0 = -1$

$t(s)$	θ (rad)	$\dot{\theta}$ (rad s^{-1})	$L\dot{\theta} + v\theta$
0	0	-12	-1.44
0.05	-0.2111	3.555	-5.903
0.075	-0.02475	11.33	0.6171
0.072	-0.05737	10.40	-0.4731
0.073	-0.04681	10.71	-0.1191
0.074	-0.03618	11.02	0.2370
0.0735	-0.04141	10.87	0.0621
0.0734	-0.04274	10.8356	0.0180
0.0733	-0.04382	10.8044	-0.0179
0.07335	-0.04328	10.8200	0.0001

From Table 11.1 we see that $(L\dot{\theta} + v\theta) = 0.0001 < 0.0005$ when $t = 0.073\,35$. So we set the new initial conditions $\theta_1 = -0.043\,28$, $\omega_1 = 10.8200$ with $t_1 = 0.073\,35$ and continue working with $\epsilon_1 = +1$.

Table 11.2 shows calculated values, correct to three significant figures, corresponding to sign changes of $(L\dot{\theta} + v\theta)$.

We note from Table 11.2 that the period of each oscillatory cycle is not constant.

$$t_2 - t_0 = 0.135\,\mathrm{s} \quad \Rightarrow \quad \text{frequency} = 7.41\,\mathrm{Hz}$$

$$t_4 - t_2 = 0.075\,s \quad \Rightarrow \quad \text{frequency} = 13.33\,\mathrm{Hz}.$$

Table 11.2 Table of values corresponding to sign changes of $(L\dot\theta + v\theta)$

$t(s)$	θ (rad)	$\dot\theta$ (rad s^{-1})	ϵ_r
$t_0 = 0$	0	-12	$\epsilon_0 = -1$
$t_1 = 0.0734$	-0.0433	10.8	$\epsilon_1 = +1$
$t_2 = 0.135$	0.0334	-8.34	$\epsilon_2 = -1$
$t_3 = 0.181$	-0.0234	5.85	$\epsilon_3 = +1$
$t_4 = 0.210$	0.0135	-3.37	$\epsilon_4 = -1$
$t_5 = 0.224$	-0.00351	0.878	

These values are to be compared with the oscillation frequency of 6–8 Hz found in practice for motorcycle front wheels.[1-3]

It will also be noticed from Table 11.2 that the amplitude θ and angular velocity $\dot\theta$ about the steering axis, are both rapidly diminishing in magnitude with increasing time.

Furthermore, when $t = t_5 = 0.224$, $\epsilon = L\dot\theta + v\theta = 0.000\,06$, and thus $\epsilon_5 = +1$; however, it is found by attempting to calculate further values that no matter how small the difference $(t - t_s)$, $L\dot\theta + v\theta$ is negative. This is a contradiction, and hence our numerical solution is valid only as far as $t = t_5$. What, then, is the subsequent motion?

Fig. 11.6 Variation of θ with t

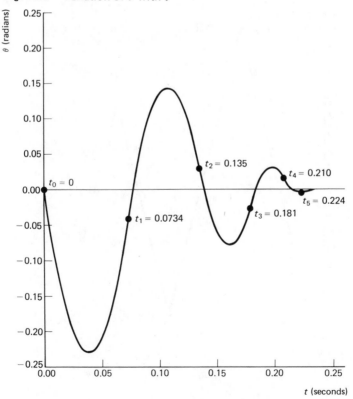

Since $\theta \neq 0$ and $\dot{\theta} \neq 0$ at $t = t_5$, then on physical grounds the motion must continue. It is clear that $L\dot{\theta} + v\theta$ is theoretically zero (0.000 06 within the accuracy of our working) when $t = t_5$. For $t > t_5$, $L\dot{\theta} + v\theta$ cannot be negative, because of the contradiction; $L\dot{\theta} + v\theta$ cannot be positive either, otherwise we would have $t_5 < 0.224$ since $\epsilon_4 = -1$. So the motion for $t > t_5$ must be described by

$$L\dot{\theta} + v\theta = 0 \qquad\qquad [11.15]$$

with initial condition

$$\theta = -0.003\,51, \quad t = 0.224 \qquad\qquad [11.16]$$

Integrating Equation [11.15] gives

$$\theta = -0.003\,51 \exp\left[-\frac{v}{L}(t-0.224)\right] \qquad\qquad [11.17]$$

and

$$\dot{\theta} = 0.003\,51\frac{v}{L}\exp\left[-\frac{v}{L}(t-0.224)\right]. \qquad\qquad [11.18]$$

Fig. 11.7 Variation of $\dot{\theta}$ with t

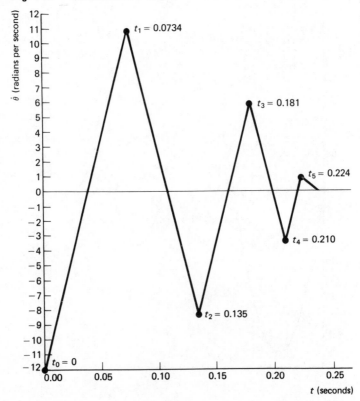

CASE STUDIES IN MATHEMATICAL MODELLING

From Equations [11.17] and [11.18], it is seen that both θ and $\dot{\theta}$ decay exponentially to zero; substituting for $v = 30\,\text{m s}^{-1}$ and $L = 0.12\,\text{m}$, then with $t = 0.300\,\text{s}$, $\theta = -2 \times 10^{-11}\,\text{rad}$, $\dot{\theta} = 5 \times 10^{-9}$ rad s^{-1}.

Figs 11.6 and 11.7 illustrate graphically the solutions to θ and $\dot{\theta}$ provided by Table 11.2 and expressions [11.17] and [11.18].

11.7 INTERPRETATION OF SOLUTION

Although the above development consists of an elementary mathematical treatment of a much simplified physical model, namely that of a castor with vertical steering axis, it is felt that considerable insight is provided into some of the aspects of oscillations in the front wheel of a motor-cycle. The frequency range calculated for the simple castor, 7–13 Hz, agrees favourably in order of magnitude with experimentally observed frequencies of 6–8 Hz in motorcycles. This agreement could well be fortuitous, of course, in view of the nature of the simplifying assumptions made in the model.

The condition $(L\dot{\theta} + v\theta) = 0$ implies that the wheel has no side-slip, whereas Equation [11.5] which refers to oscillatory motion implies that the wheel will side-slip. That a critical time has been predicted numerically for the oscillations to stop, and the subsequent motion to decay exponen-tially, requires further investigation. It can be easily shown, in general, that such a critical time exists and this is left to one of the follow-up questions in the exercise.

The mathematical analysis, as developed earlier, does not show that once $(L\dot{\theta} + v\theta) = 0$ is satisfied, the motion may partly decay exponentially and then start oscillating again (when θ and/or $\dot{\theta}$ may still be non-zero). This further analysis, which is quite straightforward and is set later as an exercise, shows that if

$$| \dot{\theta} | < \frac{\mu R L^2}{I v} \qquad\qquad [11.19]$$

is satisfied, in addition to $(L\dot{\theta} + v\theta) = 0$ (as in the numerical case considered), then the motion will decay exponentially to zero and not become oscillatory again. Naturally enough, however, if the motorcycle front wheel should be perturbed by the rider, or by a small irregularity in the road surface, then oscillations could well take place again.

A number of refinements would certainly be necessary if our mathe-matical model were to predict motorcycle speed wobble more realistically. Modelling spring tension of springs fitted on either side of the wheel axle may help in understanding the effects of tyre lateral flexibility, and introducing a spring in the steering axis might help in understanding how front wheel fork suspension affects motion. The effect of a rider holding the handlebar, thereby possibly introducing a damping couple about the steering axis, also needs further consideration. These refinements would, however, complicate the model and ensuing mathematics very considerably and probably require a quite extensive computer program to obtain solutions.

It is considered very important with this type of modelling exercise, that students should be encouraged to conduct a series of numerical experiments in solving the differential equation before attempting any analytical treatment. The 'solution', after all, can only be obtained numerically, although general analytical argument (left to the exercises) does predict a critical time at which oscillations cease and the subsequent motion decays exponentially.

11.8 RELEVANCE OF MODEL

It is clear from the way in which the model has been developed, that insights will be gained into a wide range of steering or castoring problems. The modelling treatment could also be used as an introduction to relay control theory, where the sgn function would be interpreted as a relay switching function.[2,4]

11.9 ASSESSMENT

If the model were set as an individual student exercise (perhaps as a mini-project), then it is suggested that approximately $\frac{2}{5}$ of the marks are awarded to the formulation stage, $\frac{2}{5}$ for the solution (which requires a good physical understanding of the problem), and the remaining $\frac{1}{5}$ to the interpretation of the solution (which is relatively straightforward in this case).

11.10 EXERCISES

1. Show that, for the wheel-locked case, the motion of the vertical rigid axis castor is damped simple harmonic. Use typical values: $v = 30\,\text{m s}^{-1}$, $L = 0.12\,\text{m}$, $\mu = 1$, $R = 700\,\text{N}$, $I = 0.27\,\text{kg m}^2$, and solve Equation [11.4]. Is the damping well below critical? Is it overdamped? What practical deductions can you make for a motorcyclist?

2. Using the substitutions

$$\tau = Vt/L, \quad \phi = Iv^2\theta/\mu RL^3$$

show that, for the rotating-wheel case, Equation [11.6] may be written in the dimensionless form

$$\ddot{\phi} = -\operatorname{sgn}(\dot{\phi} + \phi)$$

where differentiation is now with respect to τ.

What are the advantages of working with dimensionless forms?

3. Working with dimensionless form in Question 2, show that with the initial condition $\tau = \tau_0$, $\phi = -\omega_0$, $\dot{\phi} = \omega_0$ [and so $(\dot{\phi} + \phi)_{\tau=\tau_0} = 0$] the time increments between successive sign changes in the solution is $2[\epsilon_0\omega_0 - (2n-1)] = \tau_n - \tau_{n-1}$, for $n = 1, 2, \ldots$. Is there a value for n for which $\tau_n < \tau_{n-1}$?

Interpret physically.

4. Show that if $|\dot{\phi}| < 1$ when $\dot{\phi} + \phi = 0$, then the motion will decay exponentially to zero without subsequent oscillations.

Show that this is equivalent to

$$|\dot{\theta}| < \frac{\mu RL^2}{Iv} \quad \text{when} \quad L\dot{\theta} + v\theta = 0.$$

What happens if

$$|\dot{\theta}| = \frac{\mu RL^2}{Iv}$$

when $L\dot{\theta} + v\theta = 0$?

5. Investigate whether it is possible or not for a castor to oscillate if there is no side slip. What assumptions will you make about the frictional force acting at the point of contact of the castor with the ground? Are the predictions made by your model realistic?

Does your model relate in any way to the author's development?

REFERENCES

1. G.E. Roe 'Theory of castor oscillations', *Journal Mechanical Engineering Science* 15 (5), 379–381 (1973)
2. W.M. Pickering and D.M. Burley 'The oscillations of a simple castor', *Bulletin of the Institute of Mathematics and its Applications* 13(2), 47–50 (1977)
3. H.B. Pacejka 'Analysis of shimmy phenomenon', *Proceedings of the Institution of Mechanical Engineers* 180(2A), 251–262 (1966)
4. B.D.O. Anderson and J.B. Moore *Linear Optimal Control* (New Jersey, Prentice-Hall, 1971)
5. J.L. Synge and B.A. Griffith *Principles of Mechanics* (New York, McGraw-Hill, 1959)

12.

The News-Vendor's Dilemma

R. Croasdale

'This time last week people were clamouring to buy an evening paper and I'd sold out. So I started ordering more. Now I'm stuck with a pile of papers that nobody wants. I wish I knew how many to stock.'

(Despondent news-vendor)

12.1 STATEMENT OF THE PROBLEM

Imagine that you are taking over the running of a small kiosk retailing sweets, tobacco and the local evening paper. The paper is published as a single edition every day except Sunday. The question to be answered is 'How many copies of the paper should you order each day?' This has to be decided in advance because your order is delivered by a van which calls just once each day and you have no chance of getting more copies if you should sell out.

The printer supplies your papers at a fixed price per copy and the selling price is fixed too. There is a rebate for copies unsold, whereby some of the outlay can be recovered on copies returned to the printer.

12.2 INITIAL STUDENT RESPONSE

At this stage there should be a break to allow students ample time to consider the problem for themselves.

12.3 CONCEPTUAL APPROACH TO THE PROBLEM

In order to identify the main features of this problem, you may find it helpful to consider the following questions.

(a) What factors will affect the amount of profit you make?

(b) What factors will affect the demand and how will this vary?

(c) What should be your aim with regard to selling the paper?

(d) What data do you need to help you to determine the number of copies that you should order?

(e) How should the data be used?

Secondary Student Response

More time should be allowed here for students to consider the problem. Here are some possible responses to these questions.

(a) *Factors affecting the amount of profit*

Profit depends directly on the number of copies sold and the profit made on each of these; also on the number of copies left unsold and the loss made on each of these.

(b) *Factors affecting the demand*

Demand will vary according to current newsworthiness, e.g. topical crimes and scandals, sports news and racing results are some of the items that tend to boost sales. Sales tend to vary according to the day of the week, e.g. one day's issue may contain more advertisements of say cars or housing; Saturday's might cover football or cricket. Some days might register a lull in sales.

Some papers may be bought on the spur of the moment by people who had intended only to buy sweets or cigarettes from the kiosk.

There might be an overall tendency for sales to grow or decline.

(c) *Some possible aims*

(i) To satisfy demand.

As a news-vendor you have a responsibility both to yourself and to the buying public. You will want to be able to satisfy demand so that your customers can confidently expect to be able to buy a paper should they want one. This is in your interest, of course, because this way you will build up a regular clientèle whereas, if you are continually running out of papers, your customers will try to find a more reliable vendor.

Stocking the paper might also encourage customers to buy tobacco and sweets from you.

You should ensure that you could always satisfy demand by ordering a sufficiently large number of copies. Unfortunately you would then run the risk of having a lot of papers left unsold and these would lose money for you.

(ii) To maximise the chances of making at least £*P* profit per day where *P* is specified.

You have to make a living and presumably you will be looking for a certain return on your papers as well as on sweets and tobacco. If £*P* is the least profit you would consider to justify stocking the paper, then a strategy that maximises the chance of your making at least £*P* is reasonable.

(iii) To maximise the average daily profit in the long run.

This seems to be a reasonable aim, but note that the accent is on the long-term performance. This strategy will give varying profit because demand varies, and varies in an uncertain way, and you may have to accept quite low profits or even losses on some days in order to ensure the best return in the long run.

(d) *Data required*

You will need data so that you can assess the level of daily demand and also its variation. The Sales or Distribution Department of the paper might prove to be a useful source of information on such matters as costs and circulation figures. The total number of copies sold each day would be of some guidance, but what you would really like to know is the demand at your kiosk.

(e) *How to use the data*

If you knew in advance exactly how many papers were required each day you could order that number and so maximise your profit. Because this information is not available you have to find some way of handling the uncertainty involved.

One idea would be to find the lowest demand previously registered. If you order that number you should sell out on most days. But this might yield an unacceptably small profit since you would be missing opportunities to sell more.

At the other extreme, you could find the highest demand recorded. You know that if you order this number you will have papers left on most days — sometimes quite a lot — and that could be very unprofitable.

Alternatively, you could pick an intermediate value and order that number every day. The question is 'Which intermediate value?'.

In order to help you to answer this question, let us construct a mathematical model.

12.4 MODEL 1: HOMOGENEOUS DEMAND

We shall assume that the chances of getting a certain demand are the same on any day of any week. Mathematically speaking, we would call the corresponding demand distribution 'homogeneous'.

12.4.1 FORMULATION OF THE MODEL

From the three alternative aims listed in Section 12.3(c), we will choose aim (iii) because it is the one most likely to appeal to a rational news-vendor, i.e. we seek a policy for deciding how many copies to order each day so as to maximise the average daily profit in the long run.

It should be noted that this is not the only aim that could be adopted,[2] but this is the one that will be used throughout this account.

12.4.2 A NUMERICAL ILLUSTRATION

In order to help you to acquire a 'feel' for the situation so that you can identify the important factors in the 'general' model, the following simple example is included.

Suppose each copy you sell makes a profit of 2p and each copy left unsold loses you 3p. Tables 12.1 and 12.2 summarise the consequences of deciding to order various numbers of copies.

Table 12.1 Table showing (number sold, number unsold) for selected values of n (the number ordered) and r (the demand)

n \ r	0	1	2	3	4	5
0	(0,0)	(0,0)	(0,0)	(0,0)	(0,0)	(0,0)
1	(0,1)	(1,0)	(1,0)	(1,0)	(1,0)	(1,0)
2	(0,2)	(1,1)	(2,0)	(2,0)	(2,0)	(2,0)
3	(0,3)	(1,2)	(2,1)	(3,0)	(3,0)	(3,0)
4	(0,4)	(1,3)	(2,2)	(3,1)	(4,0)	(4,0)

Table 12.2 Table of profit (pence)

n \ r	0	1	2	3	4	5
0	0	0	0	0	0	0
1	−3	2	2	2	2	2
2	−6	−1	4	4	4	4
3	−9	−4	1	6	6	6
4	−12	−7	−2	3	8	8

If you decide to stock 2 copies, then you will make either 6p loss, 1p loss, or 4p profit, depending on whether the demand is for no papers, exactly one paper or more than one. Before you can evaluate the average daily profit resulting from this decision, you need to know the respective chances of a demand for none, for one copy and for more than one copy.

If you had decided to stock 4 copies, then you would make 12p loss, 7p loss, 2p loss, 3p profit or 8p profit depending on whether the demand is for $0, 1, 2, 3$ or more than 3 copies and the calculation of average daily profit in this case requires the respective chances of the demand being $0, 1, 2, 3$, and more than 3.

In general, then, you find that in order to calculate the average long-run daily profit, you need to know (or to be able to estimate) the probability distribution of demand.

If, for example you were to stock 4 copies and the demand for $0, 1, 2, 3$, more than 3 copies occurred in proportions 0.3, 0.2, 0.2, 0.1, 0.2, then the average daily profit would be:

$$(-12)(0.3) + (-7)(0.2) + (-2)(0.2) + (3)(0.1) + (8)(0.2) = -3.5\text{p}.$$

12.4.3 MODEL BUILDING

Suppose that you buy papers at b pence per copy and sell them at s pence per copy and that the printer allows you a pence on each unsold copy that you return.

Further, suppose you decided to stock n copies and that there is demand (or 'requests') for r copies on that day.

Now, as long as the number of copies requested is no bigger than the number of copies you stock, you will be able to satisfy all the demand, i.e.

you will sell r copies if $r \leqslant n$ and you will have $(n-r)$ left unsold.

But if the number of copies requested exceeds the number stocked, you can only sell the ones you have and some of the demand will go unsatisfied.

Then you will sell n copies if $r > n$ and you will have none left.

Now each copy sold will make you $(s-b)$ pence profit and each copy unsold will make you $(b-a)$ pence loss. So a day's profit (in pence) will be equal to

$(s-b)r - (b-a)(n-r)$ if $r \leqslant n$

(i.e. net profit from selling r copies and returning $(n-r)$ copies unsold), and

$(s-b)n$ if $r>n$

(i.e. profit from selling all n copies stocked).

Let us denote the proportion of days on which the demand is r by $f(r)$.

We now calculate your average daily profit in the long run (P_{av}, say) when you order n copies, by summing the products

(Profit) $\times$ (Proportion of days on which that amount of profit occurs)

thus:

$$P_{av} = \sum_{r=0}^{n} [(s-b)r - (b-a)(n-r)] f(r) + \sum_{r=n+1}^{\infty} (s-b)nf(r)$$

On the right-hand side of this equation, the first term in the square brackets is associated with the profit from sales when all demand is satisfied, the second term in the square brackets is associated with the loss on copies unsold when supply exceeds demand, and the third term is associated with the profit when demand exceeds supply and all stock is sold.

12.4.4   MODEL ANALYSIS

Having obtained an expression for $P_{av}(n)$, the long-run average daily profit from ordering n copies, we can write down the corresponding expression for $P_{av}(n+1)$. We can then compare the two expressions and examine whether it is more profitable to order n or $(n+1)$ copies, in the given circumstances. From this we can decide on the best value for n, the number of copies to order.

It is found (see Appendix 12.1) that the optimal value for n (n^*, say) should be chosen such that

$$\text{Prob(Demand} \leqslant n^*) < \frac{s-b}{s-a}$$

$$\text{Prob(Demand} \leqslant n^* + 1) \geqslant \frac{s-b}{s-a}$$

e.g. if you buy papers at 8p, sell them at 10p and are allowed 5p per copy returned, then

$$\frac{s-b}{s-a} = \frac{10-8}{10-5} = 0.4$$

and n^* is the 40th percentile of demand,

i.e. on 40% of days demand is less than this level, on 60% of days it is more.

In general $n*$ is the kth percentile of demand where

$$k = 100\left(\frac{s-b}{s-a}\right).$$

Note that this result depends only on the associated costs. The shape of the demand distribution (skewed, symmetrical, bi-modal or whatever) is irrelevant to the result, which is, in fact, true for any pattern of demand, provided it is homogeneous.

At first sight this might seem a surprising statement. Do you believe it? Does the result feel intuitively right? Could you have arrived at it in a simpler or more direct way?

Student Response

Time should be allowed here for students to consider these questions.

One interpretation of the solution will be familiar to students of economics, viz. the following:

When you order the optimal number of copies, the expected values of marginal cost and marginal revenue just balance. Thus, if you order one copy more than the optimal number there is a probability of 0.6 of making an extra 2p profit and a probability of 0.4 of losing a further 3p, and

$$2 \times 0.6 = 3 \times 0.4.$$

(The consequences of ordering one copy less than the optimum can be found similarly.)

When the quantity $n*$, that has been determined, is ordered, the expression for $P_{av}(n)$ simplifies to give an average daily profit of

$$P_{av}(n*) = (s-a) \sum_{r=0}^{n*} rf(r). \qquad [12.1]$$

In the absence of any real data, tutors might like to make use of the following hypothetical data:

The previous owner of the news-stand kept a careful record of the demand (i.e. the number of copies he sold or could have sold). Table 12.3 is a record of the demand over the last five weeks.

Currently the *Evening Echo* costs the retailer 8p per copy. He sells at 10p per copy and receives 5p per unsold copy returned.

Table 12.3 Demand for *Evening Echo*, weeks 1–5 (30 consecutive issues)

Week	Mon	Tue	Wed	Thu	Fri	Sat
1	203	196	213	158	176	258
2	203	226	211	150	225	248
3	211	232	210	190	225	216
4	225	229	232	200	211	240
5	223	260	219	202	197	276

Arranging the 30 values of demand in order of magnitude, starting with the smallest, then the 40th percentile of this sample lies between the 12th and 13th values, thus giving a value of $n^* = 211$ for this particular data. Alternatively, this could be obtained from a cumulative frequency polygon (or ogive).

We use this value of n^* to estimate the long-run average daily profit $P_{av}(n^*)$. To do this we substitute $s = 10$, $a = 5$ and n^* in Equation [12.1]. To evaluate the summation we need to multiply the values of r by the respective values of $f(r)$. In practice these latter values are not known so we use the relative frequencies in the sample as estimates. For the given data this will be equivalent to multiplying every observed value of demand, up to 211, by 1/30 and then summing.

Hence,

$$\text{Estimated value } P_{av}(n^*) = 5(203 + 196 + 158 + 176 + 203 + 211$$
$$+ 150 + 210 + 210 + 190 + 200 + 211$$
$$+ 202 + 196)/30$$
$$= 2717/6\,\text{p/day} = £27.17 \text{ per week.}$$

If we plot a histogram of the demand data we see that the distribution is roughly symmetrical and bell-shaped, which suggests a normal distribution.

The mean and standard deviation of demand may be estimated as 217 and 26 respectively, giving an estimate of 210 for the 40th percentile, assuming normality, which is close to the value obtained by the previous method (which was valid for any distribution).

12.4.5 VALIDATION AND MONITORING

The appropriateness of using Model 1 can only be tested over a long period of time. We have estimated the optimal n^* as best we can, given five weeks' data and we have estimated the long-run average profit using this value. We next examine the operation of our policy over the next week.

Once students have developed an order-policy, the following data and questions should be given to them. (The amount of data supplied is deliberately restricted for this exercise in order to focus attention on principles rather than arithmetic.)

Table 12.4 Demand for *Evening Echo*, week 6.

Week	Mon	Tues	Wed	Thu	Fri	Sat
6	260	234	208	174	226	269

Q.1 How much profit did you make in week 6?

Q.2 Now that you have information on six weeks' demand, what change, if any, do you wish to make to your ordering policy?

To answer the first question we calculate the profits that would result from using the order quantity previously determined. This is carried out in Table 12.5. Before answering the second question, however, you need to calculate the new value of $n*$ based on six weeks' information.

Table 12.5 Calculation of profit for week 6

Actual demand	Number ordered	Profit (p)
260	211	422
234	211	422
208	211	407
174	211	237
226	211	422
269	211	422
	Total =	£23.32

12.5 MODEL 2: DEMAND AS A TIME-SERIES

Students should now plot a time-series graph of the demand using Table 12.3 if they have not already done so.

As mentioned earlier, close inspection of the demand data from Table 12.3 indicates that the demand may be rising and that peaks and troughs tend to occur at regular intervals. This suggests that we should discard the assumption of homogeneity of the data and modify our model accordingly.

We begin by examining for trend in the data. We will illustrate the use of the semi-averages method. (Other methods for determining trend are drawing a line or simple curve 'by eye', moving averages and simple linear regression.)

We split the data into two halves, an 'early' half and a 'late' half. We calculate the mean values of demand for the two samples $\overline{Y}_E$ and $\overline{Y}_L$.

These determine two points when the mean values are plotted at the mid-points of the two time intervals. For the data in Table 12.3 we have 15 'early' values and 15 'late' values, giving $\overline{Y}_E = 208$, $\overline{Y}_L = 223$. These values are plotted at the 2nd Tuesday (i.e. day 8, the middle day of the first $2\frac{1}{2}$ week period) and the 4th Friday (i.e. day 23, the middle day of the second $2\frac{1}{2}$ week period). The join of these two points is taken as the trend line. By numbering the days starting with the first Monday of week 1 as Day 1, we can express the trend line by the equation

$$T = X + 200.$$

We now examine the data for day-to-day variations or 'daily effects'. To do this, we draw up a table of the residuals $(Y - T)$, i.e. the differences between the original values and the trend-values given by the line we have just fitted (Table 12.6).

Table 12.6 Table of residuals $(Y - T)$

Week	Mon	Tue	Wed	Thu	Fri	Sat
1	2	−6	10	−46	−29	52
2	−4	18	2	−60	14	36
3	−2	18	−5	−26	8	−2
4	6	9	11	−22	−12	16
5	−2	34	−8	−26	−32	46
Total	0	73	10	−180	−51	148
Mean	0	14.6	2.0	−36.0	−10.2	29.6
(Daily effects to nearest integer)	0	15	2	−36	−10	30

'Daily effect' here means that, for example, the average demand for papers on Tuesdays is 15 more than the number estimated by the trend line, whereas for Fridays, the average demand is 10 less than the number estimated by the trend line.

Next we examine the 'irregular' component, i.e. that which is left when we take out trend and daily variation D from the original series.

Table 12.7 Table of residuals $(I = Y - T - D)$

Week	Mon	Tue	Wed	Thu	Fri	Sat
1	2	−21	8	−10	−19	22
2	−4	3	0	−24	24	6
3	−2	3	−7	10	18	−32
4	6	−6	9	14	−2	−14
5	−2	17	−10	10	−22	16

Table 12.8 Grouped frequency table for residuals $(I = Y - T - D)$

Residual I	f	Residual I	$cf\%$
$-40-$	1	< -40.5	0.0
$-30-$	3	< -30.5	3.3
$-20-$	2	< -20.5	13.3
$-10-$	8	< -10.5	20.0
$0-$	8	< -0.5	46.7
$10-$	6	< 9.5	73.3
$20-$	2	< 19.5	93.3
$30-$	0	< 29.5	100.0
	30		

Now the analysis for Model 1 led us to choose the 40th percentile as optimal order quantity. For the current model

$$Y = T + D + I$$

where Y denotes demand, T denotes trend, D denotes the daily effect, and I denotes the residual or irregular component. The implication of this result is that we should add the 40th percentile of the I-distribution to our estimate $(T + D)$. We plot an ogive of the residuals I and use it to estimate the 40th percentile, namely -3. Using this result to estimate the optimal daily order quantities and the corresponding profit for week 6 we obtain Table 12.9.

Table 12.9 Calculation of profit for week 6 (Model 2)

	'Optimal' order quantity $(T + D - 3)$	Actual demand	Profit (p)
Mon	$231 + 0 - 3 = 228$	260	456
Tue	$232 + 15 - 3 = 244$	234	438
Wed	$233 + 2 - 3 = 232$	208	344
Thu	$234 - 36 - 3 = 195$	174	285
Fri	$235 - 10 - 3 = 222$	226	444
Sat	$236 + 30 - 3 = 263$	269	526
Total			£24.93

12.6 SIMULATION

Instead of using the data given previously, students and tutors alike might find it more interesting to make use of a routine which has been programmed in BASIC for use with a PET micro-computer (see Appendix 12.2).

The program calls on the acting news-vendor to identify a particular day of interest to him, by typing in the serial number of the day (Monday of

Week 1 has serial number 1, Tuesday of week 2 has serial number 8, etc). He then types in the number of copies he has decided to order on that day and the computer responds by displaying the demand for papers on that day and the resulting profit. The vendor then has the choice of continuing the simulation. By typing YES he can determine the profit which would result from ordering a particular number on a different day (e.g. next day), and so on until he has gathered enough information, whereupon typing NO will terminate the program. It should be noted that because of inherent variability in the demand process, repeated simulations of any given day's sales will produce differing values of demand.

This program may be used either to validate models that have been developed or to generate demand data to help with model-building.

12.7 EXTENSIONS

(a) Examine the effects on your solutions of changes in unit costs of buying, selling and returning papers.

(b) Investigate order policies designed to meet different aims.

(c) Develop a control procedure for monitoring demand so that the order quantity can respond to changes in the pattern of demand.

12.8 RELATED CASE STUDY

Tutors may like to make use of the fact that the type of reasoning utilised in this case study is similar in some ways to that required for the gas-leak problem of Chapter 14.

APPENDIX 12.1

DETERMINATION OF OPTIMAL ORDER QUANTITY n^*

The expression for $P_{av}(n)$, your long-run average daily profit if you decide to order n copies can be arranged thus:

$$P_{av}(n) = (s-b) \left\{ \sum_{r=0}^{n} rf(r) + n\, \mathrm{Prob}(r > n) \right\} - (b-a) \sum_{r=0}^{n} (n-r)f(r).$$

Correspondingly, if you decide to order $(n+1)$ copies, your long-run average daily profit $P_{av}(n+1)$ will be given by

$$P_{av}(n+1) = (s-b)\left\{ \sum_{r=0}^{n+1} rf(r) + (n+1)\,\text{Prob}(r > n+1) \right\}$$

$$-(b-a) \sum_{r=0}^{n+1} (n+1-r)f(r).$$

Having decided to try to maximise $P_{av}(n)$, you now need to know how to choose n so as to achieve this aim.

Since we are dealing with discrete values of n, we cannot use differential calculus methods. Instead, we make use of the fact that for small values of n, $P_{av}(n)$ increases as n increases and it will usually pay you to increase your stock from n to $(n+1)$ but only as long as the average profit from stocking $(n+1)$ is greater than the average profit from stocking n, i.e. as long as

$$P_{av}(n+1) > P_{av}(n)$$

or

$$P_{av}(n+1) - P_{av}(n) > 0$$

i.e.

$$(s-b)\{(n+1)f(n+1) + \text{Prob}(r > n) - (n+1)f(n+1)\}$$
$$-(b-a)\,\text{Prob}\,(r \leqslant n) > 0$$

or

$$\text{Prob}(\text{Demand} \leqslant n) < \frac{s-b}{s-a}.$$

This implies that the optimal number to order is $n = n^*$ where n^* is as shown in Fig. 12.1 or the largest integer less than n^* if n^* is not an integer.

Fig. 12.1   Ogive of demand

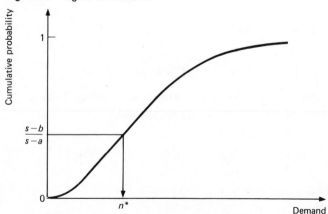

APPENDIX 12.2

PET BASIC program for News-vendor simulation

NEWSVENDOR

```
5 REM NEWSVENDOR
10 U=RND(-T1)
20 U=RND(1)
30 V=RND(1)
40 W=(SQR(LOG(1/(U↑2))))*COS(6.2832*V)
50 R=INT(20*W+.5)
60 PRINT "ENTER SERIAL NUMBER OF DAY";
70 INPUT I
80 IF I<1 THEN 370
90 PRINT"ENTER NUMBER OF COPIES ORDERED";
100 INPUT N
110 IF N<0 THEN 370
120 J=I-6*(INT(I/6))
130 PRINT "↓↓"
140 IF J=0 THEN D=20:GOTO 200
150 IF J=1 THEN D=10:GOTO 200
160 IF J=2 THEN D=0:GOTO 200
170 IF J=3 THEN D=-10:GOTO 200
180 IF J=4 THEN D=-20:GOTO 200
190 IF J=5 THEN D=0
200 Y=200+I+D+R
210 PRINT "DEMAND ON DAY";I;I;"=";Y
220 PRINT "↓↓"
230 IF Y<=N THEN 260
240 P=2*N
250 GOTO 270
260 P=5*Y-3*N
270 PRINT"PROFIT ON DAY";I;"=";P;"PENCE"
280 PRINT "↓↓↓↓↓"
290 PRINT "DO YOU WISH TO CONTINUE?"
300 INPUT A$
310 B$="YES"
320 C$="NO"
330 IF A$=B$ THEN 20
340 IF A$=C$ THEN 370
350 PRINT "PLEASE ANSWER THE NEXT QUESTION YES OR NO"
360 GOTO 290
370 STOP
READY
```

NOTE ↓ represents RVS(Q) which effects cursor down.

REFERENCES

1. M.S. Makower and E. Williamson *Teach Yourself Operational Research,* 3rd edn. (London, English Universities Press, 1973), Ch. 4, 5
2. Hong-Shiang Lau 'The Newsboy Problem under Alternative Optimization Objectives', *Journal of the Operational Research Society* 31 525–35 (1980)
3. R. Croasdale 'Statistics and the News-vendor', *International Journal of Mathematical Education in Science and Technology* 10 175–181 (1979)

13.

Modelling the True Cost of a Mortgage

M. Cross

13.1 STATEMENT OF THE PROBLEM

Imagine the following scenario some time in the mid-1970s. The manager's office of a well known building society contains three people Mr. and Mrs. Smith and the manager himself, Mr. Loansome. As we join them the following (typical) conversation is taking place:

Mr. Smith: We would like to buy a house, but don't really know much about the availability of mortgages or how much we can borrow. Can you help us?

Mr. Loansome: Since you have saved with us for some time Mr. Smith, there should be no trouble in arranging a mortgage for you. The questions you have to consider are what size mortgage you need and which type of mortgage you require.

Mr. Smith: We'd obviously like to borrow as much as possible and I'm afraid we know very little about the different types of mortgage.

Mr. Loansome: Let's deal with the amount you wish to borrow first. We normally allow married customers to borrow about two and a half times their joint income. Can you tell me your incomes?

Mr. Smith: We have a small child so my wife doesn't work and I earn around £4000 a year as a teacher.

Mr. Loansome: So that means you can borrow about £10 000 and since most mortgages are 80% of the house price, you should be looking for a house at around £12 500.

Let us now turn to the type of mortgage. Basically, there are two kinds of mortgage. In the conventional mortgage you borrow a sum at an annual interest rate (e.g. 10% at present) over a specified term, usually 25 years. We then evaluate a monthly payment in which you pay off a combination of interest and capital. Obviously, at the beginning you are paying interest whilst at the end it's mostly capital. You also have to take out a basic life insurance policy to cover the mortgage in the event of your death which costs about 35p/month/£1000 borrowed. Thus, on a £10 000 loan your gross monthly payment comes to around £100. However, to help encourage buyers the government gives full tax relief (e.g. 35%) on the mortgage interest payments and 25% on insurance policies. In the early stages this reduces your nett monthly payment to about £65.

Turning now to the second kind of mortgage — the endowment. Here you still borrow a sum of money over a specified term, but at a slightly higher interest rate (e.g. $10\frac{1}{4}$% at present). However, you don't pay back the capital directly. Instead, you pay the interest on the loan and take out an insurance policy which matures either at the end of the specified term or on the death of either of you. When the policy matures it pays off the loan and should leave around 30% of the capital as a profit. Your monthly payment, therefore, has two components, consisting of the interest on the borrowed capital plus the insurance policy at about £1.85/month/£1000 borrowed. This makes your gross monthly payment around £105 although, of course, you still get tax relief as for the conventional mortgage which reduces your nett payment to about £70.

As you can see, there are two main noticeable differences between the mortgage schemes.

(a)   the gross cost of the conventional mortgage is smaller, and

(b)   at the end of the term the endowment mortgages leaves you a nice nest egg.

Our friends may now respond as follows.

Mr. Smith: My first reaction is that the monthly payment seems quite high for both schemes though I can now see why it must be, for us to earn $10\frac{1}{2}$% interest on our savings.

Mr. Loansome: Yes, the cost appears high, but you must remember your income is likely to increase by at least the annual rate of inflation whilst the payments stay the same so that after a couple of years the cost becomes relatively small.

Mr. Smith: Yes, I suppose so. But which of these mortgages schemes is the best.

Yes — which is the best?

13.2 INITIAL STUDENT REACTION

At this stage the class should be divided into groups and dispatched to discuss the relative merits of the two mortgage schemes. In particular, they should be asked to decide upon a criterion for 'best'.

After some group discussion (for half-an-hour or so) the lecturer should then ask each group to summarise the results of their discussions. This will probably involve three main aspects — establishing an understanding of the way the mortgage schemes work, identifying the factors that influence their cost and the selection of a suitable choice criterion.

Most will soon establish all the possible factors that can influence mortgage costs (including acts of God) and will also suggest ignoring many of the more nebulous ones (e.g. moving house, non-payment, etc.) in order to reduce the difficulty of the problem. During the last decade inflation has featured prominently in most people's financial dealings and they will almost certainly wish to retain this effect.

Deciding upon the criteria for 'best' is not straightforward as there are a number of factors to consider. Based upon past experience students have come up with a number of criteria. For example,

(a) investing the difference of the cheaper mortgage each year and evaluating the returns at the end of the term,

(b) evaluating the ratio of overall return: total cost, and

(c) calculating a measure of the real cost by summing the yearly fractional costs of each mortgage (with respect to income).

During this open discussion period very little interaction with the lecturer should occur. At this stage the main task of the lecturer is merely to ensure that criterion chosen by the students is consistent and to obtain a list of agreed assumptions or simplifications.

The students may select one of the above criteria or another. In the initial model building phase this is not so important. Indeed, it only becomes important if the result/predictions it produces are unsatisfactory to the students!

13.3 ONE PARTICULAR MODEL

In the following one particular way of modelling is developed in order to assess the 'best' mortgage. It is certainly not the only way of doing so and is probably not the best. As with so much of life these days it is good enough in some nebulous sense or other.

The model building is made a good deal easier using the following (unjustified) assumptions, as a first step:

(a)   all the interest, insurance and tax rates are constant,

(b)   all payments are made on time,

(c)   no house re-sale,

(d)   no disasters (e.g. fire, etc.).

The criterion to be used in evaluating the best mortgage is that in section 13.2(c) above; that is, calculating the real cost by summing the yearly fractional cost of each mortgage (with respect to gross income) and adding any financial returns at the end of the term to the last year's income.

3.3.1   ENDOWMENT MORTGAGE

The annual payment for the endowment mortgage has a number of contributory factors which may be related as follows:

Net cost = Interest cost + endowment insurance policy
                  − tax relief on both interest and policy.

Hence, annual endowment cost,

$$C_{EA} = C\{I_e(1 - T_m) + E(1 - T_i)\}$$

where all the symbols are explained in Table 13.1.

Table 13.1   List of nomenclature

C	= Capital sum borrowed
C_{EA}	= Net annual payment on endowment
$\overline{CC}$	= Average fractional cost of conventional mortgage
$\overline{CE}$	= Average fractional cost of endowment mortgage
E	= Annual endowment insurance policy cost per £
I_c	= Conventional mortgage capital interest rate
I_e	= Endowment mortgage capital interest rate
I_s	= Annual fractional salary increase
L	= Annual conventional insurance policy cost/£
M	= Initial salary of borrower
N	= Loan term
$p(i)$	= Net annual conventional mortgage payment, year i
R_E	= Investment return from endowment policy
$S(i)$	= Capital owed after i years for conventional scheme
T_i	= Tax relief on insurance policies
T_m	= Tax relief on capital sum interest paid
x	= Gross annual conventional mortgage payment

In order to evaluate the real cost it is necessary to have some idea of the borrowers' income and how it changes during the period of the term. Since the interest rate is assumed constant, then it seems reasonable to assume that this is a realistic reflection of the inflation rate and that the borrowers income should at least keep up with inflation. Thus, if the borrower earns £M per annum initially then after r years he will earn

$$E(r) = M(1+I_s)^{r-1}.$$

Additionally, in the Nth year the borrower also receives the investment return R_E from the endowment policy. Hence, the average fractional cost of the endowment over N years is given by

$$\overline{CE} = \frac{1}{N} \left\{ \sum_{i=1}^{N-1} \frac{C_{EA}}{M(1+I_s)^{i-1}} + \frac{C_{EA}}{M(1+I_s)^{N-1}+R_E} \right\}$$

$$= \frac{C_{EA}}{NM} \left\{ \frac{(1+I_s)^{N-1}-1}{I_s(1+I_s)^{N-2}} + \frac{1}{(1+I_s)^{N-1}+R_E/M} \right\}$$

13.3.2 CONVENTIONAL MORTGAGE

In this case, the factors contributing to the net cost may be summarised as follows:

> Annual net cost = mortgage payment + insurance cost
> — tax relief on the mortgage interest and insurance.

Evaluation of the net insurance cost is similar to that for the endowment mortgage and therefore, fairly straightforward. However, this is not true of the mortgage payment. According to the manager, tax relief is only allowed on the interest paid on the outstanding capital. Thus, for each year it is necessary to evaluate not only the gross payment, but also the amount paid in interest.

Suppose, that x is the gross annual mortgage payment and $S(i)$ is the capital owed after i years. Then,

$$S(0) = C$$
$$S(1) = S(0)(1+I_c)-x$$
$$S(2) = S(1)(1+I_c)-x$$
$$= C(1+I_c)^2 - x\{1+(1+I_c)\}$$

and

$$S(i) = C(1+I_c)^i - \frac{x}{I_c}\{(1+I_c)^i-1\}.$$

Hence, if the term of the loan is N years, then $S(N) = 0$ and

$$x = \frac{CI_c(1+I_c)^N}{\{(1+I_c)^N-1\}}$$

Now the tax relief each year is on the interest paid on the capital. Therefore,

$$\text{Tax relief in year } i = I_c S(i-1) T_m.$$

Thus, the net annual payment on the conventional mortgage in year i is given by

$$p(i) = x - I_c S(i-1) T_m + LC(1 - T_i) \quad \text{for } i = 1 \text{ to } N$$

assuming for simplicity that the insurance cover is not reduced annually.

The average fractional cost over the whole term (i.e. N years) is then evaluated as for the endowment case, giving

$$
\begin{aligned}
\overline{CC} &= \frac{1}{N} \sum_{i=1}^{N} \frac{p(i)}{M(1+I_s)^{i-1}} \\
&= \frac{\{x + LC(1 - T_i)\}}{NM} \sum_{i=1}^{N} \frac{1}{(1+I_s)^{i-1}} - \frac{I_c T_m}{NM} \sum_{i=1}^{N} \frac{S(i-1)}{(1+I_s)^{i-1}} \\
&= \frac{\{x(1 - T_m) + LC(1 - T_i)\}}{NM} \sum_{i=1}^{N} (1+I_s)^{-(i-1)} \\
&\quad + \frac{T_m}{NM}(x - CI_c) \sum_{i=1}^{N} \left(\frac{1+I_c}{1+I_s}\right)^{i} \\
&= \frac{\{x(1 - T_m) + LC(1 - T_i)\}\{(1+I_s)^N - 1\}}{NMI_s(1+I_s)^{N-1}} \\
&\quad + \frac{T_m(x - CI_c)\left[\left(\dfrac{1+I_c}{1+I_s}\right)^N - 1\right]}{NM\left[\left(\dfrac{1+I_c}{1+I_s}\right) - 1\right]}
\end{aligned}
$$

13.4 RESULTS AND DISCUSSIONS

Assuming a constant inflation rate of 1% less than the conventional mortgage interest rate and an initial salary of £4000 for the borrower, then using the parameter values listed in Table 13.2 the annual net cost and the fractional cost of the endowment mortgage are

$$CEA = £832.75 \quad \text{and} \quad \overline{CE} = 0.0891$$

that is, 21% of the borrowers initial salary, but averaging out at around 9% of the annual salary over the term of the loan.

Table 13.2   Data used in calculations

C	£10 000
E	0.0222 £/£/yr
I_c	10 %/yr
I_e	9 %/yr
L	0.0042 £/£/yr
M	£4000/yr
N	25 years
R_E	£3000
T_i	25 %/yr
T_m	35 %/yr

Again, using the data summarised in Table 13.2, the annual gross mortgage payment, the net payment in the first year and the fractional cost of the conventional mortgage may easily be evaluated as

$$x = £1101.68 \qquad p(1) = £783.18 \quad \text{and} \quad \overline{CC} = 0.08999.$$

Comparing the above results for each of the mortgage schemes it is clear that their fractional cost averaged over the loan term is almost identical. Moreover, although the initial cost of the endowment policy is rather more than a conventional mortgage, its return to the customer in real terms at the end of the loan is rather small. This is clear when the endowment return of £3000 is compared with the borrowers salary at the end of the term (approximately £34 000).

Thus, it would appear that, in general, the conventional mortgage is the 'best'. Finally, according to this model the idea of investment associated with the endowment scheme is almost 'mythical'. All that is happening in this case is that the borrower pays the insurance company fairly dearly to provide him with a severely devalued sum at the end of the mortgage term. Should we, therefore, be surprised that insurance companies do so well!

13.5   SUGGESTION FOR FURTHER WORK

As it stands, the present model assumes that all the interest and tax rates stay constant. Initially, it would be interesting to evaluate how sensitive the model results (and, therefore, the conclusions) are to different tax and mortgage interest rates. Thus, a sensitivity analysis could prove invaluable.

Secondly, the assumption that all the model parameters remain constant throughout the term does not match reality. As such, it could prove interesting to develop a model to cope with annual variations in any of the parameters and then reperform the above analysis.

14.

Searching for Gas Leaks

A.E. Millington

14.1 STATEMENT OF THE PROBLEM

The following situation is one which is known to occur. A slight gas leak exists in a domestic gas supply between the meter and the various gas appliances in the house. The pipe above ground is inspected and no leak is detected. One or two floorboards are removed in a further attempt to locate the leak below ground, but without success. At this stage either more floorboards, etc. must be removed in the hope that the leak will be found, or the search abandoned and complete new piping installed.

The problem, therefore, is as follows. A leak exists in a length of pipe, and must be mended either by locating it and repairing or by replacing the piping. The cost of repairing a leak, once located, is small, but it is by no means certain that a leak which does exist in a piece of exposed piping will actually be located by the usual test method. (A typical success rate is 90%, and somewhat less for underground pipe.) In other words, even if every centimetre of pipe is exposed and tested, there is still a chance that no leak will be located.

In these circumstances the Gas Board management sees it appropriate to lay down general guidelines to service engineers in searching for leaks, and in particular indicating the stage at which the search should be abandoned.

14.2 ADVICE TO THE LECTURER

Once the students have been presented with the statement of the problem, they should be left to discuss it in order to identify the main factors involved. The problem here hinges on the difference between the two types of pipe. Not only is that below ground more difficult and therefore more expensive to expose and test, but also the chance of locating a leak in that exposed piping is somewhat less than for pipe above ground. The statement of the problem does not stress this, and while a competent class may be expected to reason it out themselves, the lecturer may decide to assist a less experienced class by providing the data available which highlight the point. In either case the students should be able to isolate the main parameters as listed in Section 14.3 and appreciate the need for the data contained in Section 14.4.

The aim is to minimise the cost to the customer by deciding when to stop searching. The suggested model assumes that:

(a)   the pipe above ground is searched before that below ground,

(b)   we aim to minimise *expected* cost.

The statement of the problem hints at (a) and this is also what happens in practice.

The choice of (b) may not be so obvious. The objection that this will not guarantee the best result for a particular customer may be raised, or some students may wish to minimise the maximum cost. Such points are valid, and a discussion on the relative merits of expected value would not be out of place.

If we calculate the total expected cost *given* the (maximum) lengths of pipe to be searched then a slightly different model emerges, but the results obtained on minimising this cost are identical with those presented here.

14.3   POLICY

One criterion for ending the search is to stop once the amount of saving expected in a further search becomes negative, i.e. stop when the cost of replacement will be, on average, less than the total cost of continuing the search. This suggests a stochastic approach to the problem.

The main factor in this problem is the difference between the two types of piping, namely that above ground and that below. The latter is far more expensive to test, and far more difficult to test with a high degree of efficiency. For this reason it is sensible not to test any underground pipe until all that above ground has been (unsuccessfully) tested. The following variables are therefore critical:

L_a,   the length of pipe above ground.

L_b,   the length of pipe below ground.

$L(= L_a + L_b)$,   the total length of pipe.

γ_a,   the cost of searching unit length of pipe above ground.

γ_b,   the cost of searching unit length of pipe below ground.

γ,   the cost of repairing leak.

C,   the total replacement cost (dependent on L_a, L_b).

π_a,   the success rate of finding leaks in exposed pipe above ground, i.e.

   π_a = Prob(find leak in length l/leak exists in l) and is independent of l, $0 < l < L_a$.

π_b,   the success rate of finding leaks in exposed pipe below ground, similar to π_a.

14.4   TYPICAL DATA

The following data are available.

Of 20 m of piping about one third is above ground, and the estimated replacement cost is £135. The cost of repairing a leak varies from £1 to £5. The success rate for detecting faults in pipe above ground is about 0.9 whereas that for pipe below ground can vary from 0.4 to 0.7 depending on the age of the pipe and construction of the house. The cost per metre of searching pipe above ground is estimated to be £1.30 and for pipe below ground between £2.50 and £6.00.

14.5   FORMULATION

Since it is relatively cheap to inspect above-ground pipe and there is a good chance of detecting any leak that occurs above ground, it would seem to be a good idea to inspect all above-ground pipe first. Assuming that this is our policy our problem divides into two cases, viz searching above ground and then searching below ground.

14.5.1   SEARCHING ABOVE GROUND

Let us consider a stage at which we have searched a length l without detecting the leak. We have to decide whether or not to continue the search to the next bit of pipe, length Δl, say. If we decide to abandon the search at this stage it will necessitate full replacement of the pipe, at cost C.

If we decide to inspect this additional length Δl, we incur the cost $\gamma_a \Delta l$ of inspection (whatever the outcome), plus **either** γ, the cost of repair, (in the event of our finding a leak) **or** C, the cost of replacing the pipe (if we do not find a leak).

For the length Δl, the probability of finding a leak given the leak was not found in length l is p, and the probability of not finding a leak given the leak was not found in length l is $(1-p)$.

The decision process may be represented in the form of a decision tree (Fig. 14.1) and the expected cost may be used as a criterion for decision making.

Fig. 14.1 Decision tree for inspection of length Δl above ground

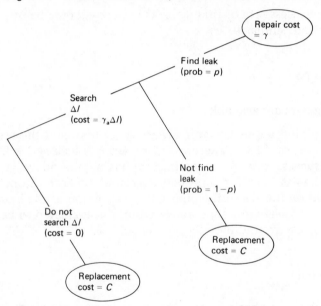

The expected cost of deciding to inspect Δl is $\gamma_a \Delta l + p\gamma + (1-p)C$ whilst the expected cost of deciding not to inspect Δl is C. It is worthwhile, therefore, to continue inspecting at l if

$$C > \gamma_a \Delta l + p\gamma + (1-p)C$$

that is, if

$$p(C-\gamma) - \gamma_a \Delta l > 0. \qquad [14.1]$$

Now let us assume that the leak occurs randomly in the pipe, that is, a length Δl in one part of the pipe is just as likely to contain a leak as a length Δl in any other part of the pipe. Then the probability that the leak is in the first l units of pipe is l/L whilst the probability that the leak is not in the first l units of pipe is $(1-l/L)$.

Now we know that if we have not found a leak in the first l units then, **either** the leak is not in the first l units (the probability of this is $1-l/L$) **or** the leak is in the first l units and we have not found it (the probability of this is $(l/L)(1-\pi_a)$).

Thus,

$$\text{Prob(leak not found in } l) = (1-l/L)+(l/L)(1-\pi_a) = 1-(l/L)\pi_a$$

and

$$p = \text{Prob(leak is found in } \Delta l \mid \text{not found in } l)$$

$$= \frac{\text{Prob(leak not found in } l \text{ and is found in } \Delta l)}{\text{Prob(leak not found in } l)}$$

$$= \frac{\text{Prob(leak is found in } \Delta l)}{\text{Prob(leak not found in } l)}$$

as we assume there is only one leak,

$$= \frac{\pi_a(\Delta l/L)}{1-(l/L)\pi_a} = \frac{\pi_a \Delta l}{L - \pi_a l}.$$

Inequality [14.1] now becomes

$$\frac{\pi_a \Delta l}{L - \pi_a l}(C-\gamma)-\gamma_a \Delta l > 0$$

which is equivalent to the following.

Continue inspecting at length l if

$$l > L/\pi_a - (C-\gamma)/\gamma_a. \qquad [14.2]$$

Now define $\theta_a = L/\pi_a - (C-\gamma)/\gamma_a$. If $\theta_a < 0$ then inequality [14.2] will always hold, since $l \geqslant 0$, and it follows that at *any* stage in our search above ground we can expect to save money by inspecting more pipe. In particular, to minimise the average total cost, we should be prepared to inspect all the pipe above ground.

(If $\theta_a > 0$ then our policy is not so clear. In the long run we certainly expect to lose by being prepared to search a length less than θ_a. But it may be that if we are prepared to search all of the pipe above ground then we would expect to pay less than if we searched none.)

14.5.2 SEARCHING BELOW GROUND

In accordance with our policy we only consider a search of the pipe below ground if we have unsuccessfully inspected the entire piping above ground.

Let us assume that we have inspected a length l of pipe below ground, and we are deciding whether or not to search a further small length Δl. Let P be the conditional probability of finding the leak in Δl given that we have not found it so far (in $L_a + l$). The argument is similar to that of Section 14.5.1.

The expected cost of deciding to inspect Δl is

$$\gamma_b \Delta l + P\gamma + (1-P)C$$

and the expected cost of deciding not to inspect Δl is C, so that we continue searching if

$$P(C-\gamma) - \gamma_b \Delta l > 0. \qquad [14.3]$$

The leak is in length L_a with probability L_a/L and in length l with probability l/L. If we have not found it then **either** it is not in the length $L_a + l$ (with probability $1-(L_a + l)/L$) **or** it is in length L_a and we have not found it (probability $(1-\pi_a)L_a/L$) **or** it is in length l and we have not found it (probability $(1-\pi_b)l/L$).

Hence

Prob(leak not found in $L_a + l$)

$$= 1-(L_a + l)/L + (1-\pi_a)L_a/L + (1-\pi_b)l/L$$

$$= 1-(\pi_b l + \pi_a L_a)/L$$

so that

$$
\begin{aligned}
P &= \frac{\text{Prob(leak not found in } L_a + l \text{ and leak found in } \Delta l)}{\text{Prob(leak not found in } L_a + l)} \\[2mm]
&= \frac{\text{Prob(leak is found in } \Delta l)}{\text{Prob(leak not found in } L_a + l)} \\[2mm]
&= \frac{\pi_b(\Delta l/L)}{1-(\pi_b l + \pi_a L_a)/L} \\[2mm]
&= \frac{\pi_b \Delta l}{L - \pi_b l - \pi_a L_a}
\end{aligned}
$$

and inequality [14.3] becomes

$$\frac{\pi_b \Delta l}{L - \pi_b l - \pi_a L_a}(C-\gamma) - \gamma_b \Delta l > 0.$$

Therefore, it is certainly worthwhile continuing the search if

$$l > \theta_b \qquad\qquad\qquad\qquad\qquad [14.4]$$

where

$$\theta_b = \frac{L}{\pi_b} - \frac{\pi_a L_a}{\pi_b} - \frac{(C - \gamma)}{\gamma_b}.$$

If $\theta_b < 0$ it is always to our advantage to search more pipe, and our best policy would be to search all of the pipe below ground. If $\theta_b > 0$ then Equation [14.4] will hold only after a certain length of pipe, θ_b, has been tested. We expect to lose by being prepared to search a limited length of pipe, but if we are prepared to inspect the entire length of pipe then this may be our best policy.

14.6 DISCUSSION OF THE MODEL WITH THE DATA GIVEN

We will first consider the piping above ground. Putting $L = 20$, $\pi_a = 0.9$, $C = 135$, $\gamma = 3$, $\gamma_a = 1.3$ gives $\theta_a = -79.3 < 0$.

This means that we should be prepared to inspect all the pipe above ground, and that at any stage in the search we reduce our expected cost by inspecting more pipe.

For the pipe below ground putting $L = 20$, $\pi_a = 0.9$, $\pi_b = 0.7$, $C = 135$, $\gamma = 3$, $\gamma_b = 2.5$ gives $\theta_b = -32.8 < 0$.

In this case, where the pipe is relatively cheap to inspect, and the success rate is high our best policy is to inspect as much pipe as possible, and we reduce our expected cost at any stage by testing more pipe.

Putting $L = 20$, $\pi_a = 0.9$, $\pi_b = 0.4$, $C = 135$, $\gamma = 3$, $\gamma_b = 6.0$ gives $\theta_b = 13 > 0$. Notice that in this case $\theta_b \simeq L_b$ so, where the pipe is very expensive to test and the success rate is very low, it pays us to search none of the pipe, since at any stage in such a search we expect to lose money by continuing.

In between these two extremes it may be the case that if we are prepared to search all the pipe, then we expect to reduce our cost. To investigate this we need to calculate the *expected* costs of being prepared to search:

(a)   all the pipe above but none below, and

(b)   all the pipe above and below.

(a)   The leak is above ground with probability L_a/L. Given that it is above ground we find it with probability π_a having searched, on average, length $L_a/2$, and on average we incur a cost of $\gamma_a L_a/2 + \gamma$. We fail to find it with probability $(1 - \pi_a)$ and in this case we search the whole length L_a and then have to pay the replacement cost, the total being $\gamma_a L_a + C$. The leak is below ground with probability L_b/L. Given that it is below ground we would search length L_a and then pay the replacement cost of C. On average our cost would be $\gamma_a L_a + C$.

The total expected cost is therefore

$$\frac{L_a}{L}\left[\pi_a\left(\frac{\gamma_a L_a}{2}+\gamma\right)+(1-\pi_a)(\gamma_a L_a + C)\right]+\frac{L_b}{L}(\gamma_a L_a + C)$$

$$= C + \frac{L_a \gamma_a \pi_a}{L}\left(\theta_a - \frac{L_a}{2}\right).$$

(b) The leak occurs above ground with probability L_a/L. Given that it is above ground we find it with probability π_a having searched, on average, $L_a/2$, and repair it. We fail to find it with probability $(1-\pi_a)$ and we search a total length of $L_a + L_b$ and then replace the pipe.

The leak occurs below ground with probability L_b/L. Given that it is below ground we find it with probability π_b having inspected, on average, $L_a + L_b/2$, and repair it. We fail to find it with probability $(1-\pi_b)$ having searched $L_a + L_b$ and we then replace the pipe.

The total expected cost is

$$\frac{L_a}{L}\left[\pi_a\left(\gamma_a \frac{L_a}{2}+\gamma\right)+(1-\pi_a)(\gamma_a L_a + \gamma_b L_b + C)\right]$$

$$+\frac{L_b}{L}\left[\gamma_a L_a + \pi_b\left(\gamma_b \frac{L_b}{2}+\gamma\right)+(1-\pi_b)(\gamma_b L_b + C)\right]$$

$$= C + \frac{L_a \gamma_a \pi_a}{L}\left(\theta_a - \frac{L_a}{2}\right)+\frac{L_b \gamma_b \pi_b}{L}\left(\theta_b - \frac{L_b}{2}\right).$$

The above argument is illustrated diagramatically in Fig. 14.2.

The second policy is better if and only if

$$\frac{L_b \gamma_b \pi_b}{L}\left(\theta_b - \frac{L_b}{2}\right)< 0$$

i.e. if

$$\theta_b < \frac{L_b}{2}$$

i.e. if

$$\frac{L}{\pi_b} - \frac{(C-\gamma)}{\gamma_b} - \frac{\pi_a}{\pi_b} L_a < \frac{L_b}{2}.$$

The following rule thus emerges.

(a) If $\theta_b > L_b/2$ then there is no point in searching any of the pipe below ground.

(b) If $0 < \theta_b < L_b/2$ then our best policy is to be prepared to search all of the pipe below ground.

(c) If $\theta_b < 0$ then any search below ground is advantageous while the best policy is to search all.

Fig. 14.2 Decision tree for search above and below ground

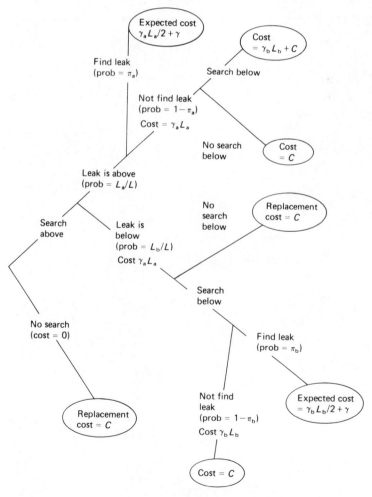

For example, for the case in which $L = 20$, $\pi_a = 0.9$, $\pi_b = 0.5$, $\gamma_b = 5.0$, $C = 135$ and $\gamma = 3$, $\theta_b = 1.6 < 6.7$ so that by (c) it is worthwhile to inspect the pipe below ground, if we are prepared to inspect it at all.

14.7 EXTENSIONS TO THE MODEL

The above discussion has only considered one set of values of C, L_a and L_b. An extension would be to consider how the decision would change for a different ratio L_a/L, remembering that this will generally mean a change in the value of C too.

Other extensions would be to consider either a problem with more than one leak, or more than two levels of pipe.

15.

Selling Hot Dogs

A. Rothery

15.1 STATEMENT OF THE PROBLEM

At a college a hot-dog stall operates over a half-hour period during the lunch interval. The stall operator wishes to run his stall in the most profitable manner and to this end he has amassed the following details which he considers to have some bearing on the situation.

The rolls used for the hot-dogs cost 50p a dozen, the sausages cost 60p a pound (8 sausages), and the onions about 30p per pound. Each hot-dog needs one roll, one sausage and $\frac{1}{2}$ oz of onions. In addition about £1's worth of sauce is used over a lunch time. Rolls left over are bought by the staff at 2p each.

The cooking is done using gas cylinders and about £8s worth of fuel is used over a 5-day week. The cost of equipping the stall is about £5 per week.

Laundry charges amount to about 80p for each person employed plus £1.40 for towels, etc.

It takes about 25 to 35 seconds for an assistant to serve a customer depending on whether or not the customer requires change. It is thought that about 50% of customers give the correct money.

During the last week 624 customers were served and 823 hot-dogs were sold at a price of 30p each. As usual, surplus cooked sausages were sold to the staff at 5p each and total takings for the week (including sales of surplus rolls) came to £251.33.

Although the stall is open only half-an-hour each day, staff employed have to work $1\frac{1}{2}$ hours each day so that preparation and cleaning may be done. The hourly rate of pay for staff is £1.50 and staff are paid when on sick leave.

The data of Table 15.1 relate to the arrival pattern at the stall and were collected by students doing a statistical project. In the course of data collection it was also noted that arrivals tended not to join the queue if there were seven or more customers already there.

Find the best way of organising the stall — particularly the best number of servers to be employed.

Table 15.1 Arrival pattern at stall

Number of arrivals during a 30 second period	Frequency: number of 30 second periods in which the given number of arrivals was observed
0	49
1	26
2	18
3	10
4	10
5	8
6	7
7	6
8	5
9	4
10	3
11	2

15.2 TACKLING THE PROBLEM

The statement of the problem is rather detailed and this is typical of many problems in economics. One of the main aims in discussing problems of this type is to help students distinguish between key factors and others of lesser importance. There is probably no right way of simplifying the problem, though some simplifications may be sillier than others. The main objective for the tutor is to ensure that each student constructs a feasible simplified problem from the set problem.

In developing a model we attempt, where appropriate, to replace the given factual data by variables and parameters. In this way we create a generalised model which can be used to solve not only the given problem but a related class of decision-making problems.

15.3 MODELLING THE NUMBER OF CUSTOMERS SERVED

The data of Table 15.1 is developed to display the total number of arrivals as in Table 15.2.

Table 15.2 Number and frequency of arrivals at stall

Number of arrivals during half a minute	Frequency: number of half-minute periods in which the given number of arrivals was observed	Number of arrivals
0	49	0
1	26	26
2	18	36
3	10	30
4	10	40
5	8	40
6	7	42
7	6	42
8	5	40
9	4	36
10	3	30
11	2	22
	148 (i.e. 74 mins.)	384

From Table 15.2 it is easily seen that the mean arrival rate is $384/74 = 5.2$ people per minute.

5.3.1 AN ALGEBRAIC MODEL

A simple model for the service capacity:

$$\left(\begin{array}{c}\text{The service}\\ \text{capacity}\end{array}\right) = \frac{\left(\begin{array}{c}\text{Length of time}\\ \text{stall is open}\end{array}\right)\left(\begin{array}{c}\text{Number of}\\ \text{servers}\end{array}\right)}{\text{Service time}}$$

where we have made the further assumption that the serving time is constant.

We can use this model to examine the effect on the service capacity of changing the number of servers. For example, in a 30 second serving time

$$\text{The service capacity in } \tfrac{1}{2} \text{ hour} = \begin{cases} 120 \text{ customers with 2 servers} \\ 180 \text{ customers with 3 servers.} \end{cases}$$

A simple model for the arrival of customers is:

$$\left(\begin{array}{c}\text{The number of}\\ \text{arrivals}\end{array}\right) = \left(\begin{array}{c}\text{Mean arrival}\\ \text{rate}\end{array}\right)\left(\begin{array}{c}\text{Length of time}\\ \text{stall is open}\end{array}\right).$$

For the data of Table 15.2 this model predicts 156 arrivals during a half-hour period.

The conclusion from these two simple models is that with two servers

120 customers are served and 36 are turned away, whilst with three servers, all 156 get served.

More generally, the number of customers, N, served would be

$$N = \begin{cases} rT & \text{if } r \leqslant n/\tau \quad \text{(i.e. arrival rate less than or equal to the service rate)} \\ \dfrac{Tn}{\tau} & \text{if } r > n/\tau \quad \text{(i.e. arrival rate greater than service rate)} \end{cases}$$

where T is the length of time the stall is open, r is the mean arrival rate, n is the number of servers, and τ is the time taken to serve one customer.

Assuming, for example, that the mean arrival rate is 5.2 people per minute, the serving time is 30 seconds and the stall is open for half-an-hour, then in this half-hour, the total number of customers served is

$$N = \begin{cases} 156 & \text{if } n \geqslant 2.6 \\ 60n & \text{if } n < 2.6. \end{cases}$$

The obvious interpretation of this result is that, with only 1 or 2 servers, not all customers arriving would get served. Three servers would seem to be better but we would have to look carefully at the trade-off between increased income and the cost of employing an additional server. This is taken up in Section 15.4. Clearly, however, more than three servers would be pointless.

15.3.2 A SIMULATION MODEL

A deficiency of the simple algebraic model in Section 15.3.1 is that the arrival of customers is unlikely to be regular. There could, therefore, be times when the servers are idle and hence the number of people served in half-an-hour would be less than the estimate given by the simple model. One way of handling the situation is to use a simulation approach; that is, we attempt to generate a sequence of customers which has the characteristics of the data in Table 15.1. A simple version of this is to take a 'top hat' simulation of half-an-hour's business using the data of Table 15.1. For simplicity, we again assume that the service time is constant.

A 'top hat' simulation for the given data involves taking 148 small cards and putting '0' on 45 of them, '1' on 26 of them, '2' on 18 of them, '3' on 14 of them, and so on. The cards are then placed in a suitable container and drawn at random, replacing the cards after the draw. Each draw represents the number of arrivals in the simulation, as the probability of choosing a certain number of arrivals is thus matched to the observed frequency of arrivals.

Using this method of simulating the number of arrivals each half-minute, a record can be kept of the number served and the numbers currently in the queue. In an example simulation, 168 arrivals were produced, and the numbers served were calculated from the arrival pattern (a) with two servers and (b) with three servers. In both cases, the number served was also calculated on the assumption that the arrivals go away if the queue exceeds 7. The results were:

Arrivals: 168		
2 servers:	117 served	(allowing queues to be long)
	112 served	(if customers are turned away on seeing a queue over 7 people long)
3 servers:	147 served	(allowing long queues)
	139 served	(restricting queues to 7)

The result here showed that with 3 servers the effect of 'idle time' cut down the amount of business which could be done. The models in Section 15.3.1 predicted that 3 servers would serve *all* possible arrivals, whereas the simulation showed that 3 servers would *not* be able to serve all arrivals.

A better simulation technique would be to find a probability distribution based on the given data.

Rather than use a 'top hat', a relatively simple computer program could generate the arrivals using random numbers. This could then be run and re-run several times; and could carry out investigation of various numbers of servers, and serving times.

Clearly the effect of 'idle time' (i.e. time when, by chance, no customers have arrived and the servers are idle) is noticeable. Furthermore the off-putting effect of long queues (which crop up at random) is significant. Both these factors need to be investigated in further simulation work.

For students with a knowledge of queuing theory, an 'algebraic' analysis is, of course, feasible.

15.4 INVESTIGATING THE PROFIT EXPECTED

The decision whether to employ two or three servers depends on the profit £P. The following model is now quite obvious.

$$P = L - \frac{3nw}{2} - H$$

where P is the profit in the half-hour, L is the net income in the half-hour, n is the number of servers, w is the hourly wages of a server, and H is the overheads for half-an-hour.

In using the above model we will consider the net income L to be the income from the sale of the hot dogs, less the cost of materials. In H we will include the cost of gas, towels, laundry, etc. From the data, one element of H, laundry costs, depends on the number of servers, n, but this element is so insignificant we will proceed by considering H to be independent of n. We have also ignored the income from surplus sold to staff.

15.4.1 NEGLECTING IDLE TIME

Using the algebraic model of Section 15.1.1 for the number of customers served, we can consider the profit P_2 for two servers and P_3 for 3 servers. These are

$$P_2 = 120c - 3w - H$$

and

$$P_3 = 156c - \frac{9w}{2} - H$$

where £c is the average net income per customer.

Thus $P_2 - P_3 = (3w/2) - 36c$ and this means that $P_2 > P_3$ if $w > 24c$.

Hence, if the hourly wage of a server is greater than 24 times the net income per customer, just two servers would be employed, otherwise three servers would be employed.

From the data given in Section 15.1,

$$C = \frac{823}{624} \times \left(30 - \frac{50}{12} - \frac{60}{8} - \frac{30}{32}\right) p = 22.95 \mathrm{p}$$

and it is clear, from the above analysis, that, in this case, three servers should be employed. Also, there would have to be very significant changes in wages and/or prices and/or costs for this policy to be changed. It is left as an exercise for the student to evaluate the profit under this policy and to examine the sensitivity of the policy to changes in wages, prices and costs.

15.4.2 ALLOWING FOR IDLE TIME

Using a similar approach but, this time with the results of Section 15.3.2, we obtain

$$P_2 = 112c - 3w - H$$

and

$$P_3 = 139c - \frac{9w}{2} - H.$$

Thus $P_2 > P_3$ if $w > 18c$ and this still indicates, that, with current wages, etc., three servers should be employed.

15.5 CONCLUSIONS AND FURTHER INVESTIGATION

The approach taken so far has been dominated by the investigation of how many servers to employ. To achieve this there was considerable simplification of the situation and a certain amount of given data was ignored. However, there is a clear recommendation to use three servers, though the investigation can be developed along the following lines.

(a) Allow for more of the given information: laundry charges, re-sale of left-over rolls, etc.

(b) Carry out a more sophisticated and wide-ranging simulation exercise to build a more accurate picture of how many hot dogs would be sold with three servers; this for instance can further extend the work on 'idle time' and the loss of custom due to long queues.

(c) Consider the expected profit more precisely, i.e. by estimating overheads, etc.

There are also further avenues for exploration in the development of particular problems. For example: what happens if one server is ill?

16.

Product Pricing

M.A. Wilson

16.1 STATEMENT OF THE PROBLEM

C. Threwes Ltd. is a small expanding firm that manufactures delicate transparent objects. The firm is organised as four main sections which are the office, the glass shop, the plastics shop, and 'stores'. Details of what is done by each section have been obtained from discussion with the heads of each of the sections and are summarised as follows.

The glass shop employs twenty people at an average wage of £100 a week. Having made an item, a craftsman then records the time used for that job, on a job ticket. Material costs are negligible in comparison with wages. Each employee works a forty-hour week but only about twenty-five hours of this appears on job tickets since staff tend not to record times used for design and material selection which takes the form of discussion with workmates and customers.

The plastics shop holds ten expensive machines and thirty people are employed there. The average wage is also £100 a week. As with the glass shop, craftsmen log time spent on jobs on the job tickets. This recorded time accounts for about 63% of their 40-hour week. Again material costs can be neglected.

The 'stores' are somewhat mis-named in that they cover a number of functions. These include security, cleaning, machine maintenance, and the obtaining and storing of raw materials. Their weekly wage bill for the twenty-two staff (which includes part-time workers) is about £750. The section estimates that about 200 man-hours a week are spent on problems directly associated with production, this work being evenly split between the glass and plastics shop. In addition, about 50 man-hours work is needed for tasks directly associated with the office, including regular polishing of the Sales Director's car!

The office also covers a number of functions. One of these, namely accounts, is poorly performed. It is unwieldy due to the company growth and is run by a man who is rather inept in this field. The office estimate that they spend 210 man-hours a week on work directly associated with the production departments — about two thirds of this for the plastics shop if one operates on order volumes. Work for 'stores' is also estimated at about 140 man-hours each week. (These estimates are vague as they had never been required before.) The weekly wage bill (plus director's fees) is £1050 in round terms.

As far as can be determined fixed costs are about £320 a week for each section (office apart) and in addition, stores use about £80's worth of fuel, oil and petrol each week. The costs for the office, which include many miscellaneous items, amount to £560 per week. There are also die rentals and depreciation to consider for the plastics' machines and these work out at £1100 a week.

At the moment the price of an item is worked out by taking the hours recorded on its job ticket and charging this at £8.50 per hour. There is a pay rise due that will affect the office and about 50% of the 'stores' staff. This will increase the wage bill in those sections by 12% and 5% respectively.

Taking, as an example, a job ticket with four hours glass shop work recorded on it, then what did it cost the firm overall to make, and what effect will the pay rise have on this cost? What would you suggest as a fair price for the item?

16.2 INITIAL REACTION

The project is devised for students with little background in accountancy (though even accountants may benefit from it!). Because of the difficulty students experience in sorting out the problem the outline situation contains little information that is unnecessary.

Groups tackling the problem initially encounter three problem areas:

(a) getting to grips with the data,
(b) introducing deliberate omissions such as staff absences,
(c) perceiving how to disperse office and 'stores' costs.

In addition, debate occurs about fairness.

Initial assistance could be directed to summarising the situation in a succinct way. Thereafter two directions can be investigated; either an arithmetic apportioning of the costs over the other departments, or charging an hourly rate for services so that income balances expenditure for these sections. It has been observed that most students start with the apportioning idea and for that reason the model developed starts from that position.

16.3 DEVELOPMENT OF A MODEL

The situation can be readily summarised as in Tables 16.1 and 16.2.

Table 16.1 Weekly summary (production)

Department	Glass Shop	Plastics Shop
Capacity (man-hours)	500	750
Wages	£2000	£3000
Fixed cost	£320	£1420
Total cost	£2320	£4420

Table 16.2 Weekly summary (services to production)

Department	Stores	Office
Capacity (man-hours)	250	350
Wages	£750	£1050
Fixed cost	£400	£560
Total cost	£1150	£1610

Each service section uses its capacity in providing the following man-hours of service to the other sections as indicated in Table 16.3.

Table 16.3 Allocation of service

	Stores	Office	Glass	Plastics
Stores	—	50	100	100
Office	140	—	70	140

For convenience we shall assume that each unit of product requires one man-hour's work and shall term glass, product A and plastic, product B. For the coming week we expect to make 500 A's and 750 B's. If we sell these at prices of £p and £q per unit respectively then the revenue obtained, namely £$(500p + 750q)$, must exceed our total cost of £9500; that is

$$2p + 3q > 38.$$

Any values of p and q satisfying this (as in the current situation) will do for the purposes of making a profit. Any further search is for values of p and q that are in some way fair. We take this to mean values that reflect the resources used in the manufacture of the products.

If we assume that each hour of service provided by Stores is equally valuable (rather unlikely in the real situation) then one idea that can be tried is to allocate the total cost of Stores to the other sections in proportion to the man-hours provided to those sections (see Table 16.3), appropriately re-allocating any charges made to Stores by other sections providing services to Stores. In this case, for Stores, 20% of its costs go to the Office, 40% to each of the work shops. Similarly, for the Office, 40% is allocated to each of Stores and the Plastics Shop and 20% to the Glass Shop. The arithmetic is shown in Table 16.4, single barline denoting totals and double underlining indicating the amount being shared out on the next line.

Table 16.4 Allocation table

Stores	Office		Glass	Plastics	
(-/20/40/40)	(40/-/20/40)				
1150	1610				
	230		460	460	(= 1150)
		($\overline{\overline{1840}}$)			
736			368	736	(= 1840)
	147.2		294.4	294.4	
58.9			29.4	58.9	
	11.8		23.6	23.6	
4.7			2.4	4.7	
	0.9		1.9	1.9	
0.4			0.1	0.4	
			0.2	0.2	
1950*	2000*		1180	1580	

Such an allocation shows that the Glass Shop uses service resources worth £1180 and the Plastics Shop uses services worth £1580. We note that the total value of resources provided £(1150 + 1610) equals the total cost allocated £(1180 + 1580).

By tabulating as in Table 16.5 we are in a position to answer the first question posed obtaining minimum values for p and q as 7 and 8 respectively. One would expect some increase on these values to obtain a profit. If we say a 12% profit (as now) is needed then $p = 7.84$ and $q = 8.96$; under such a scheme the appropriate charge for the order would be £31.36. Wage rises may be dealt with in similar fashion but a search for methods to avoid repetitive arithmetic could be sought at this stage. To achieve this we wish to arrive at the distribution of service costs by a more revealing method than the above arithmetic.

Table 16.5 Cost per unit

	Glass Shop	Plastics Shop
Wages	2000	3000
Fixed costs	320	1420
Services	1180	1580
Total	3500	6000
Cost per unit	3500/500 = 7	6000/750 = 8

Ignoring the two starred values for the moment, we observe that 40% of each element under Stores in Table 16.4 is entered in the Glass Shop column as is 20% of each value under the Office column. Hence the total for the Glass Shop is 40% of the Stores total plus 20% of the Office total; similarly the Plastics total is 40% of the Stores total plus 40% of the Office total. If we term these totals under Stores and Office as X and Y then the apportioned costs become

$$0.4X + 0.2Y \quad \text{for the Glass Shop} \tag{16.1}$$

and

$$0.4X + 0.4Y \quad \text{for the Plastics Shop.} \tag{16.2}$$

If we look again under the Stores column of Table 16.4 we note that X is the sum of 1150 plus 40% of each element under the Office column and similarly Y is 1610 plus 20% of each element under the Stores column; or formally

$$X = 1150 + 0.4Y \tag{16.3}$$

$$Y = 1610 + 0.2X. \tag{16.4}$$

Solving Equations [16.3] and [16.4] for X and Y must give the starred values 1950 and 2000 shown in Table 16.4 and then apportioning by Equations [16.1] and [16.2] must result in 1180 and 1580 as being the apportioned costs for services. This can easily be verified.

For want of a better term, the values of X and Y will be given the description 'conceptual' costs in the rest of these suggestions. Working in such a manner allows us to deal with changes and possible alternatives more rapidly.

16.5 GENERALISED MODEL

Denote the given values, in the original situation, in the tabular format of Tables 16.6 and 16.7.

Table 16.6 Service provided

Hours service provided	To / By	Stores	Office	Glass	Plastics	Total
	Stores	0	τ_{12}	t_{11}	t_{12}	T_1
	Office	τ_{21}	0	t_{21}	t_{22}	T_2

Table 16.7 Section costs

Section	Stores	Office	Glass	Plastics
Wages	W_1	W_2	C_1	C_2
Fixed cost	F_1	F_2	R_1	R_2

Introducing vector-matrix notation, let

$$W = (W_1 \ \ W_2), \quad C = (C_1 \ \ C_2), \quad F = (F_1 \ \ F_2), \quad R = (R_1 \ \ R_2)$$

$$\alpha = \begin{bmatrix} \alpha_{11} & \alpha_{12} \\ \alpha_{21} & \alpha_{22} \end{bmatrix} \qquad \beta = \begin{bmatrix} \beta_{11} & \beta_{12} \\ \beta_{21} & \beta_{22} \end{bmatrix}$$

where $\tau_{ij}/T_i = \alpha_{ij}$ $(\alpha_{ii} = 0)$ and $t_{ij}/T_i = \beta_{ij}$.

We then have that the conceptual cost vector $E = (E_1 \ \ E_2)$ is given by

$$E = F + W - E\alpha$$

which, on rearranging, gives

$$E = (F + W)(I - \alpha)^{-1}$$

where I is the 2×2 identity matrix.

The allocation vector $S = (S_1 \ \ S_2)$ is given by

$$S = E\beta$$

$$= (F + W)(I - \alpha)^{-1}\beta, \quad \text{by above.}$$

Thus, the totals matrix $P = (P_1 \ \ P_2)$ is given by

$$P = C + R + S$$

$$= C + R + (F + W)(I - \alpha)^{-1}\beta$$

which enables us to find the total cost for each production section.

Increasing vector and matrix dimensions enables us to deal with the question of several production and service departments.

16.6 SPECIAL NOTE

If such a system were used in practice then it would probably take the form of applying overheads to the manufacturing time. Taking the Glass Shop for example,

Wages	2000
Fixed cost	320
Services	1180
Total	3500
÷ (Capacity × Wage rate)	3500/1250 = 2.8
Overheads	$(2.8 - 1) \times 100 = 180\%$
Time used 4 hours, labour cost	£10
add 180% overheads	£18
Total	£28

which is in agreement with the earlier zero profit situation.

16.7 EXTENSIONS

In practice, capacity (man-hours) is estimated in advance. Should demand for the product be low, it may not all be utilized. If demand is high then it can be expanded by the use of overtime, which will result in hourly wage costs, in the production departments, being increased (say by 30%). Assuming demand fluctuates by 5% from the given capacity figures, will this necessitate any change of view on the price to be used?